Springer Proceedings in Physics

Volume 298

Indexed by Scopus

The series Springer Proceedings in Physics, founded in 1984, is devoted to timely reports of state-of-the-art developments in physics and related sciences. Typically based on material presented at conferences, workshops and similar scientific meetings, volumes published in this series will constitute a comprehensive up to date source of reference on a field or subfield of relevance in contemporary physics. Proposals must include the following:

- Name, place and date of the scientific meeting
- A link to the committees (local organization, international advisors etc.)
- Scientific description of the meeting
- List of invited/plenary speakers
- An estimate of the planned proceedings book parameters (number of pages/articles, requested number of bulk copies, submission deadline).

Please contact:

For Americas and Europe: Dr. Zachary Evenson; zachary.evenson@springer.com
For Asia, Australia and New Zealand: Dr. Loyola DSilva; loyola.dsilva@springer.com

Ri-ichi Murakami · Mikito Yasuzawa ·
Yoshinobu Shimamura · Pankaj Koinkar ·
Hairus Abdullah · Antonio Nakagaito
Editors

The 3rd International Conference on Nanomaterials and Advanced Composites

Proceedings of NAC 2022, July 15–17, Tokushima, Japan

Springer

Editors
Ri-ichi Murakami
Chengdu University
Chengdu, China

Yoshinobu Shimamura
Department of Mechanical Engineering
Shizuoka University
Hamamatsu, Shizuoka, Japan

Hairus Abdullah
Department of Materials Science and Engineering
National Taiwan University of Science and Technology
Taipei, Taiwan

Mikito Yasuzawa
Graduate School of Technology, Industrial and Social Sciences
Tokushima University
Tokushima, Japan

Pankaj Koinkar
Graduate School of Technology, Industrial and Social Sciences
Tokushima University
Tokushima, Japan

Antonio Nakagaito
Graduate School of Technology, Industrial and Social Sciences
Tokushima University
Tokushima, Japan

ISSN 0930-8989 ISSN 1867-4941 (electronic)
Springer Proceedings in Physics
ISBN 978-981-99-7152-7 ISBN 978-981-99-7153-4 (eBook)
https://doi.org/10.1007/978-981-99-7153-4

This Springer imprint is published by the registered company Springer Nature Singapore Pte Ltd.
The registered company address is: 152 Beach Road, #21-01/04 Gateway East, Singapore 189721, Singapore

Paper in this product is recyclable.

Participating Countries and Organizations

Chairman

Mikito Yasuzawa, Tokushima University, Japan

Co-chairmen

Yoshinobu Shimamura, Shizuoka University, Japan
Yun-Hae Kim, Korea Maritime and Ocean University, Republic of Korea
Chang-Mou Wu, National Taiwan University of Science and Technology, Taiwan, R. O. C.

International Organizing Committee

Ri-ichi Murakami, Chengdu University, China
Tae-Gyu Kim, Pusan National University, Republic of Korea
Seung-Hyo Lee, Korea Maritime and Ocean University, Republic of Korea
Wen-Cheng Ke, National Taiwan University of Science and Technology, Taiwan, R. O. C.
Keh-Moh Lin, Southern Taiwan University of Science and Technology, Taiwan, R. O. C.
Mahendra More, Savitribai Phule Pune University, India
Subhash Kondawar, Rashtrasant Tukadoji Maharaj Nagpur University, India
Jun-Cai Sun, Dalian Maritime University, China
Dongyan Zhang, Xidian University, China
Toshihiro Moriga, Tokushima University, Japan
Hitoshi Takagi, Tokushima University, Japan

Pankaj Koinkar, Tokushima University, Japan
Antonio Norio Nakagaito, Tokushima University, Japan
Masashi Kurashina, Tokushima University, Japan
Tomoyuki Fujii, Shizuoka University, Japan
Tomoki Yabutani, Ehime University, Japan
Yusuke Fuchiwaki, National Institute of Advanced Industrial Science and Technology (AIST) Shikoku, Japan

Local Organizing Committee

Mikito Yasuzawa, Tokushima University, Japan
Toshihiro Moriga, Tokushima University, Japan
Hitoshi Takagi, Tokushima University, Japan
Pankaj Koinkar, Tokushima University, Japan
Antonio Norio Nakagaito, Tokushima University, Japan
Masatsugu Oishi, Tokushima University, Japan
Yoshihisa Suzuki, Tokushima University, Japan
Kei-ichiro Murai, Tokushima University, Japan
Masashi Kurashina, Tokushima University, Japan
Yoshinobu Shimamura, Shizuoka University, Japan
Tomoyuki Fujii, Shizuoka University, Japan
Tomoki Yabutani, Ehime University, Japan
Yohei Yamada, National Institute of Technology, Anan College, Japan

Participating Countries

1. India
2. Israel
3. Japan
4. P. R. China
5. Republic of Korea
6. Taiwan (R. O. C.)

Preface

We are delighted to present this issue in the book series Springer Proceedings in Physics containing 10 selected papers out of talks from four countries that were presented at the 3rd International Conference on Nanomaterials and Advanced Composites (NAC 2022) held at Tokushima University, Japan, from July 15th to 17th of 2022. The very first NAC conference started in Busan, South Korea in 2018, which served as an event to bring together researchers and engineers from academia and industry for the purpose of exchanging updated knowledge on diverse fields spanning nanotechnology, nanomaterials, and advanced composites. The second conference, which took place in Taipei, Taiwan in 2019, was a more expansive event but still maintained the same objectives initially established. However, due to the global pandemic, the third conference was conducted in hybrid mode by combining on-site and online presentations. The majority of the attendees joined remotely from their respective countries. But again, the initial aims to allow participants to share the most recent information was not forgotten, and networking of academics and professionals in the concerned fields was encouraged, despite the fact that it was done remotely. About 90 participants were involved in this conference. The speakers, professors, and oral/poster presenters from various universities in India, Israel, Japan, P. R. China, Republic of Korea, and Taiwan (R. O. C.) actively presented their research works. The overall organization of the conference had one public event session, four keynote sessions, 14 regular sessions (featuring 15 invited talks and 48 oral presentations), and one poster session. We have high hopes that this special issue will not only serve as a representative record for the event but also provide new ideas for future studies.

Chengdu, China — Prof. Ri-ichi Murakami
Tokushima, Japan — Prof. Mikito Yasuzawa
Hamamatsu, Japan — Prof. Yoshinobu Shimamura
Tokushima, Japan — Dr. Pankaj Koinkar
Taipei, Taiwan — Dr. Hairus Abdullah
Tokushima, Japan — Dr. Antonio Nakagaito

About This Book

This book presents selected articles from the 3rd International Conference on Nanomaterials and Advanced Composites (NAC 2022) held at Tokushima University in Japan. This event brought together leading researchers and professionals from academia and industry to present their latest findings and served as a platform for the exchange of ideas aiming further collaborations.

Participants from over six countries shared their most up-to-date knowledge in their respective fields covering nanotechnology, nanomaterials, and advanced composites. Even though this conference had both on-site and remotely connected attendees, the main purpose to promote the networking among academics, engineers, and students was fully achieved. This book is part of the effort to disseminate the knowledge gathered during this meeting.

The collection of articles covers topics on advanced composites, nanomaterials, ecological materials, energy, microfluidics, crystal growth, and photocatalysis. This representative account of the conference intends to provide new and useful insights for prospective studies in materials science and engineering.

Contents

Part I Advanced Composites and Eco-Friendly Materials Technology

1 **Damage Behavior of Carbon/Epoxy Laminated Composites Composed of Super-Thin Plies** 3
Cheng-Yu Chen and Wen-Shyong Kuo

2 **Tensile Properties of Fiber-Reinforced Plastic-Based Epoxy Prepregs Storable at Room Temperature** 9
Yoh Kataoka, Katsushige Kouge, and Koichi Goda

3 **Effect of Process Parameters on Feasibility of Production of Cellulose Nanofiber Yarn by Wet Spinning** 23
Yoshinobu Shimamura, Hiroki Kondo, and Tomoyuki Fujii

Part II Energy and Sensor Materials and Technology

4 **Synthesis of *N*-Methyl-D-Glucamine Modified Chitosan Nanofibers for Boron Adsorption** 31
Masashi Kurashina, Daiki Kato, Haoyuan Li, Keita Shiba, Yuta Morishita, Kazuki Shibata, Ho Hong Quyen, and Mikito Yasuzawa

5 **Practical Microfluidic Technologies for In-Vitro Diagnostic Devices** 37
Yusuke Fuchiwaki

Part III Advanced Nanomaterials and Nanotechnology

6 ***In-Situ* Growth of Silicon Nanowires Array and Its Field Emission Behavior** 47
Amol B. Deore, Krishna Jagtap, Sandesh R. Jadkar, and Mahendra A. More

7 **Photoluminescence Property of Nano Silica Mixed YAG:Ce Phosphors** ... 57
Tatsuki Sogabe, Takaaki Sakai, Satoshi Hiroi, Koji Ohara, Satoshi Sugano, Shao-Ju Shih, Toshihiro Moriga, and Masatsugu Oishi

8 **In Situ Observation of Crystal Growth Processes** ... 67
Yoshihisa Suzuki, Ai Ninomiya, and Shinichiro Yanagiya

9 **Approach for Achieving Effective Photocatalytic Activity Under Visible Light of WO_{3-x}/SnO_2 Produced by Laser Ablation Method** ... 75
Vinayak Shinde, Tetsuro Katayama, Yasuyuki Maeda, Satoshi Sugano, Akihiro Furube, and Pankaj Koinkar

10 **Study on Cellulose Nanofiber Molding by 3D Printing** ... 85
Yuta Yokota, Antonio Norio Nakagaito, and Hitoshi Takagi

Contributors

Cheng-Yu Chen Xitun Dist, Feng Chia University, Taichung City, Taiwan

Amol B. Deore Centre for Advanced Studies in Materials Science and Condensed Matter Physics, Department of Physics, Savitribai Phule Pune University, Pune, India

Yusuke Fuchiwaki Health and Medical Research Institute, AIST, Takamatsu, Kagawa, Japan;
New-Generation Medical Treatment and Diagnosis Research Laboratory, AIST, Tsukuba, Ibaraki, Japan;
Advanced Photonics and Biosensing Open Innovation Laboratory, AIST-Osaka Unverisity, Photonics Center, Osaka University, Suita, Osaka, Japan

Tomoyuki Fujii Shizuoka University, Hamamatsu, Japan

Akihiro Furube Institute of Post-LED Photonics and Graduate School of Technology, Industrial and Social Sciences, Tokushima University, Tokushima, Japan

Koichi Goda Department of Mechanical Engineering, Yamaguchi University, Tokiwadai, Ube, Japan

Satoshi Hiroi Diffraction and Scattering Division, Japan Synchrotron Radiation Research Institute (JASRI), Hyogo, Japan

Sandesh R. Jadkar Centre for Advanced Studies in Materials Science and Condensed Matter Physics, Department of Physics, Savitribai Phule Pune University, Pune, India

Krishna Jagtap Centre for Advanced Studies in Materials Science and Condensed Matter Physics, Department of Physics, Savitribai Phule Pune University, Pune, India

Yoh Kataoka Paper Industry Innovation Center, Ehime University, Shikokuchuo, Japan

Tetsuro Katayama Institute of Post-LED Photonics and Graduate School of Technology, Industrial and Social Sciences, Tokushima University, Tokushima, Japan

Daiki Kato Department of Applied Chemistry, Graduate School of Science and Technology, Tokushima University, Tokushima, Japan

Pankaj Koinkar Institute of Post-LED Photonics and Graduate School of Technology, Industrial and Social Sciences, Tokushima University, Tokushima, Japan

Hiroki Kondo Graduate School of Integrated Science and Technology, Shizuoka University, Hamamatsu, Japan

Katsushige Kouge Sanshin Chemical Industry Co., Ltd, Minamihama, Yanai, Japan

Wen-Shyong Kuo Xitun Dist, Feng Chia University, Taichung City, Taiwan

Masashi Kurashina Department of Applied Chemistry, Graduate School of Science and Technology, Tokushima University, Tokushima, Japan

Haoyuan Li Department of Applied Chemistry, Graduate School of Science and Technology, Tokushima University, Tokushima, Japan

Yasuyuki Maeda Department of Optical Science, Graduate School of Technology, Industrial and Social Sciences, Tokushima University, Tokushima, Japan

Mahendra A. More Centre for Advanced Studies in Materials Science and Condensed Matter Physics, Department of Physics, Savitribai Phule Pune University, Pune, India

Toshihiro Moriga Graduate School of Technology, Industrial and Social Sciences, Tokushima University, Tokushima, Japan

Yuta Morishita Department of Applied Chemistry, Graduate School of Science and Technology, Tokushima University, Tokushima, Japan

Antonio Norio Nakagaito Graduate School of Technology, Industrial and Social Sciences, Tokushima University, Tokushima, Japan

Ai Ninomiya Graduate School of Technology, Industrial and Social Sciences, Tokushima University, Tokushima, Japan

Koji Ohara Diffraction and Scattering Division, Japan Synchrotron Radiation Research Institute (JASRI), Hyogo, Japan

Masatsugu Oishi Graduate School of Technology, Industrial and Social Sciences, Tokushima University, Tokushima, Japan

Ho Hong Quyen Faculty of Environment, University of Science and Technology, The University of Da Nang, Da Nang, Vietnam

Takaaki Sakai Global Zero Emission Research Center, National Institute of Advanced Industrial Science and Technology (AIST), Tsukuba, Ibaraki, Japan

Keita Shiba Department of Applied Chemistry, Graduate School of Science and Technology, Tokushima University, Tokushima, Japan

Kazuki Shibata Department of Applied Chemistry, Graduate School of Science and Technology, Tokushima University, Tokushima, Japan

Shao-Ju Shih Department of Materials Science and Engineering, National Taiwan University of Science and Technology, Taipei, Taiwan

Yoshinobu Shimamura Shizuoka University, Hamamatsu, Japan

Vinayak Shinde Department of Optical Science, Graduate School of Technology, Industrial and Social Sciences, Tokushima University, Tokushima, Japan

Tatsuki Sogabe Graduate School of Technology, Industrial and Social Sciences, Tokushima University, Tokushima, Japan

Satoshi Sugano Graduate School of Technology, Industrial and Social Sciences, Tokushima University, Tokushima, Japan

Yoshihisa Suzuki Graduate School of Technology, Industrial and Social Sciences, Tokushima University, Tokushima, Japan

Hitoshi Takagi Graduate School of Technology, Industrial and Social Sciences, Tokushima University, Tokushima, Japan

Shinichiro Yanagiya Institute of Post-LED Photonics, Tokushima University, Tokushima, Japan

Mikito Yasuzawa Department of Applied Chemistry, Graduate School of Science and Technology, Tokushima University, Tokushima, Japan

Yuta Yokota Graduate School of Advanced Technology and Science, Tokushima University, Tokushima, Japan

Part I
Advanced Composites and Eco-Friendly Materials Technology

Chapter 1
Damage Behavior of Carbon/Epoxy Laminated Composites Composed of Super-Thin Plies

Cheng-Yu Chen and Wen-Shyong Kuo

Abstract According to past research, using thinner layers usually results in a stronger composite. Along with the size effect, thinner layers also lead to a larger interfacial area, which enhances load transfer. In this work, a super-thin carbon/epoxy prepreg was utilized for making laminated composites. The prepreg thickness was 23 microns, which is about five-times thinner than conventional layers. Specimens of cross-ply laminates were prepared and cured by compression molding. Two types of specimen panels were made, composed of 50 and 60 layers of the super-thin ply. For comparison purposes, specimens composed of 8-layer standard plies were made. All these specimens have the same carbon fiber and epoxy resin, and their fiber contents are very close. Three-point bending and compressive tests were conducted. Scanning electron microscopy (SEM) was used to observe the induced damage. The results indicate that the properties were notably improved by using the super-thin plies. The fracture behavior and induced damage modes were very different. Some unique damage patterns were observed, which were unseen in composites composed of standard plies.

1.1 Introduction

It is known that when a material becomes thinner, its strength is generally improved [1, 2]. This is so-called the size effect. A thinner material is usually accompanied with fewer and smaller defects, which reduce the stress intensity and are less detrimental when subjected to external loads. The same concept can be applied to the composite plies. In recent years, using thinner plies to replace the conventional ones in laminated

C.-Y. Chen (✉) · W.-S. Kuo
Xitun Dist, Feng Chia University, No. 100, Wenhua Rd, Taichung City, Taiwan
e-mail: j8281015@gmail.com

W.-S. Kuo
e-mail: wskuo@fcu.edu.tw

R. Murakami et al. (eds.), *The 3rd International Conference on Nanomaterials and Advanced Composites*, Springer Proceedings in Physics 298,
https://doi.org/10.1007/978-981-99-7153-4_1

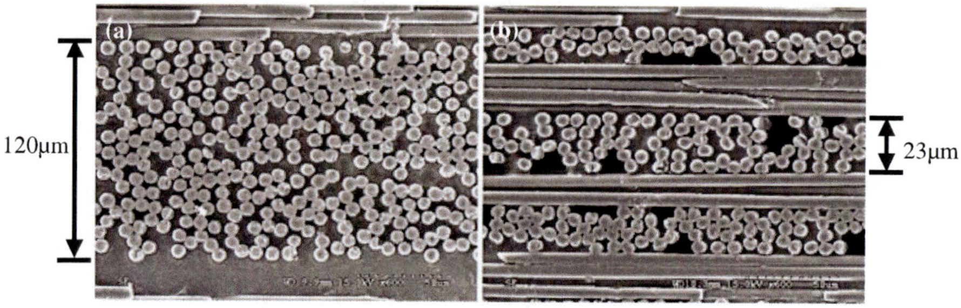

Fig. 1.1 SEM images of present specimens composed of **a** standard plies and **b** super-thin plies (magnification × 600)

composites has received significant attention. Making thinner plies requires a technology called spread-tow. A spread-tow leads to a thinner thickness in the resulting prepreg. The through-thickness resin impregnation becomes easier and fiber wetting is better [3]. In addition, another important advantage is the increased total interfacial area in the composite, which enhances load transfer between layers and the capability of energy absorption when impacted [4–6]. In the literature, most research on laminated composites were based upon standard plies with thickness in the range of 100 ~ 120 microns [7]. Some thin-ply research were based on ply-thickness of 50 ~ 60 microns. In this research, the 23-micron ssuper-thin prepreg was used. Figure 1.1 compares the present samples composed of conventional 120-microns and 23-microns plies, respectively.

1.2 Experimental Method

1.2.1 Materials and Lamination Design

The 23-microns prepreg was provided by TYKO Tech of Taiwan. Both the standard prepreg and the super-thin prepreg were used to make cross-ply laminates. Table 1.1 shows the specifications of the prepregs. There are three types of laminates in this research, as shown in Table 1.2. Both the STD8 and Thin50 have the nominal thickness of 1.0 mm. The STD8 and Thin60 have the same total FAW (fiber areal weight) of 1200 g/m^2 for the laminates.

Table 1.1 Prepreg specifications

Prepreg type	Carbon fiber level	RC%	FAW (g/m^2)	Thickness (mm)
Standard	Toray T700	38	150	120
Super-thin	Toray T700	36	20	23

Table 1.2 Specimens and ply-stacking

Prepreg type	Specimen	Layup	Layer number	Thickness (mm)	FAW (g/m^2)
Standard	STD8	$[0/90]_4$	8	1.0	1200
Super-thin	Thin50	$[0/90]_{25}$	50	1.0	1000
Super-thin	Thin60	$[0/90]_{30}$	60	1.2	1200

1.2.2 Fabrication of Composite Laminates and Testing

In this experiment, the carbon composite laminates were made by compression molding. The curing temperature was set at 150 °C for 30 min, and the pressure was maintained at 10 kg/cm^2 during the hot pressing. The cured laminates were then cut and ground to the desired sizes for the tests.

The three-point bending tests were conducted according to the ASTM D790-10. The specimen sizes were 50 mm in length and 10 mm in width. The span-thickness ratio in the test was 32:1, and the pressing speed was 0.5 mm/min. The compressive test was based on the ASTM D6641-09 standard. The specimen sizes were 45 mm in length and 10 mm in width, and the pressing speed was 0.5 mm/min.

1.3 Results and Discussion

Typical results of the loading curves of the bending and compressive tests are shown in Figs. 1.2a and 1.2b, respectively. All the material specimens show brittle fracture in the bending tests. The curves are linear up to the point of fracture. The slopes of the Thin50 and Thin60 are close, both noticeably higher than the STD8. This means the flexural stiffness values are higher. The highest points are also much higher than the STD8, indicating the higher strength and toughness in the thin-ply specimens. The stress–strain curves of the compressive tests showed some small drops in the curve, revealing that minor damage modes occur before the final fracture. Both the Thin50 and Thin60 are higher than the STD8 in the compressive strength and the failure strain.

The results of the 3-point tests are shown in Fig. 1.3, based on the same-thickness condition. Compared with the STD8 specimen, the Thin50 had the flexural strength and modulus increased by 33.3% and 50.5%, respectively. The compressive strength and modulus were increased by 44.3% and 28.0%, respectively. If the comparison was made based on the same total FAW, the Thin60 specimens had flexural strength and modulus 42.6% and 58.5% higher. The compressive strength and modulus were increased by 57.5% and 46.7%, respectively.

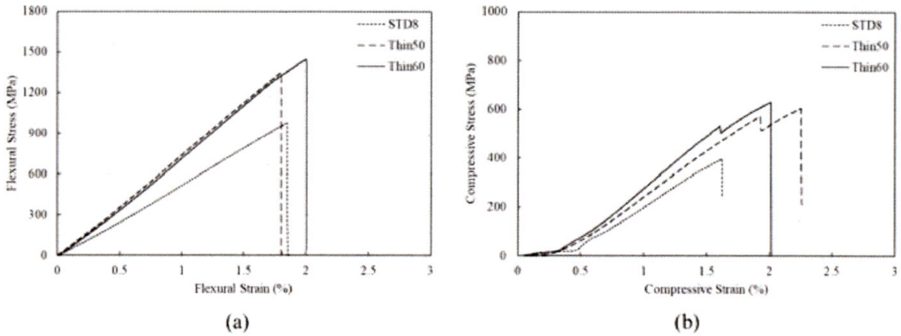

Fig. 1.2 **a** Typical loading curves of the flexural tests and **b** the compressive tests

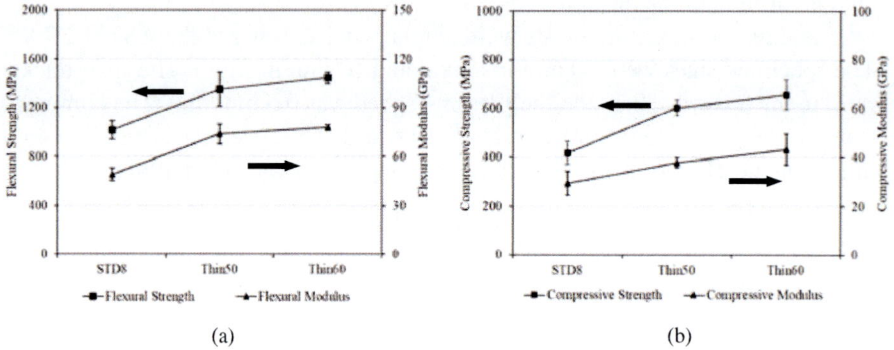

Fig. 1.3 Comparison of **a** the obtained flexural strength and modulus; **b** the compressive strength and modulus

Figure 1.4 shows typical fracture in the specimens after the bending test. The cracks start from the tensile side (the bottom) and grow upward toward the top. Because the layer numbers are higher, the Thin50 and Thin60 specimens reveal more rugged crack growth. Crack deflection is more noticeable, meaning that the total crack surface is increased, and more fracture energy is consumed. Both lead to slower crack growth, and a larger load is needed to break the specimen.

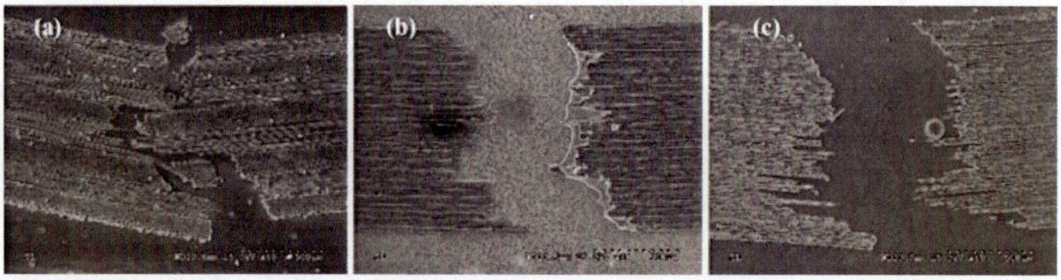

Fig. 1.4 Side views of the fractured specimens after the flexural tests. **a** STD specimen, **b** Thin50 specimen, **c** Thin60 specimen. (magnification $\times$ 60)

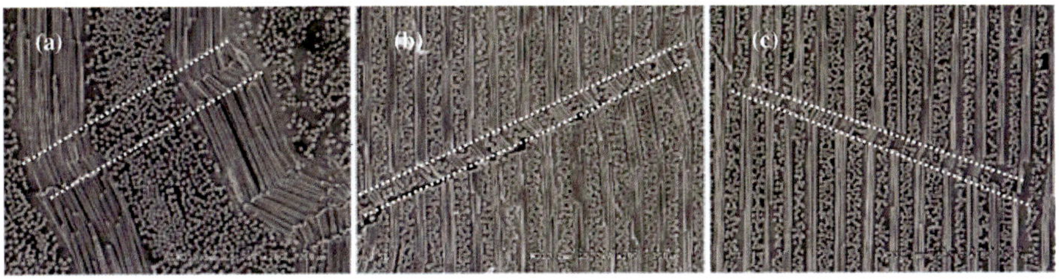

Fig. 1.5 SEM pictures of the compression specimens showing the kink-bands of fractured fibers. **a** STD, **b** Thin50, **c** Thin60. (magnification × 200)

The major damage mechanism in the compressive test is the formation of fiber kink-band, resulting from fiber buckling, as shown in Fig. 1.5. The widths of all observed kink-bands were measured. The results show that the kink-band width was in the range of 70 ~ 80 microns for the STD8, while the width was 38 ~ 50 microns for the super-thin specimens. The narrower kink-band width is due to the thinner plies and is the key to increase the compressive strength.

1.4 Conclusions

This work studied the effect of using super-thin layers in the laminated composite. The thinner thickness leads to a higher interfacial area, which provides a more effective load transfer. The bending specimens show more rugged fractured surfaces. Both flexural modulus and strength are significantly increased. In the compressive test, fiber kink-band is the major mode of damage. Using the super-thin prepreg leads to shorter kink-band widths and a higher compressive strength.

Acknowledgements The authors thank TYKO Tech Co., Ltd. for providing the material and the Ministry of Science and Technology of Taiwan (grant no. MOST 109-2622-8-035-001-TE4).

References

1. Parvizi, A., Garrett, K.W., Bailey, J.E.: Constrained cracking in glass fibre-reinforced epoxy cross-ply laminates. J. Mater. Sci. **13**, 195–201 (1978)
2. Sihn, S., Kim, R.Y., Kawabe, K., Tsai, S.W.: Experimental studies of thin-ply laminated composites. Compos. Sci. Technol. **67**, 996–1008 (2007)
3. Amacher, R., Cugnoni, J., Botsis, J., Sorensen, L., Smith, W., Dransfeld, C.: Thin ply composites: Experimental characterization and modeling of size-effects. Compos. Sci. Technol. **101**, 121–132 (2014)
4. Arteiro, A., Catalanotti, G., Xavier, J., Camanho, P.P.: Notched response of non-crimp fabric thin-ply laminates. Compos. Sci. Technol. **79**, 97–114 (2013)
5. Arteiro, A., Catalanotti, G., Xavier, J., Camanho, P.P.: Large damage capability of non-crimp fabric thin-ply laminates. Compos. A **63**, 110–122 (2014)

6. Cugnoni, J., Amacher, R., Kohler, S., Brunner, J., Kramer, E., Dransfeld, C., Smith, W., Scobbie, K., Sorensen, L., Botsis, J.: Towards aerospace grade thin-ply composites: Effect of ply thickness, fibre, matrix and interlayer toughening on strength and damage tolerance. Compos. Sci. Technol. **168**, 467–477 (2018)
7. Mazumdar, S.K.: Composites manufacturing, materials, product and process engineering, 1st edn. CRC Press, Boca Raton (2002)

Chapter 2
Tensile Properties of Fiber-Reinforced Plastic-Based Epoxy Prepregs Storable at Room Temperature

Yoh Kataoka, Katsushige Kouge, and Koichi Goda

Abstract Carbon fiber-reinforced plastics (CFRPs) are widely used in various parts of the engineering field. Carbon fibers are among the materials with the highest specific strengths that are commonly applied in fiber-reinforced plastics. In many cases, epoxy resins are utilized as CFRP matrices, which can exhibit various physical properties, such as adhesiveness and heat resistance, depending on the curing agent. These resins are also attracting significant attention in the field of composite materials owing to their high strength and low weight. Composite materials containing epoxy resins are often stored in the form of prepregs at low temperatures (generally below 0 °C). If the mass-produced prepregs could be stored at room temperature, it would eliminate the need for bulky equipment (such as freezers) and considerably reduce manufacturing costs and environmental burden (including energy consumption). Consequently, many currently used transportation methods and conventional manufacturing processes would be changed dramatically. Thus, the purpose of this study is to investigate the possibility of using a pre-cured resin storable at room temperature as a CFRP matrix material.

2.1 Introduction

Thermoplastic resins, which require short molding times, are increasingly used as matrices for the mass production of fiber-reinforced plastics (FRPs), which exhibit superior specific strength and specific stiffness. However, thermoplastics generally possess lower mechanical strength and heat resistance than those of thermosetting

Y. Kataoka (✉)
Paper Industry Innovation Center, Ehime University, Otsu 127, Mendori-Cho, Shikokuchuo, Japan
e-mail: kataoka.yo.bg@ehime-u.ac.jp

K. Kouge
Sanshin Chemical Industry Co., Ltd, 4-4-6, Minamihama, Yanai, Japan

K. Goda
Department of Mechanical Engineering, Yamaguchi University, 2-16-1, Tokiwadai, Ube, Japan

R. Murakami et al. (eds.), *The 3rd International Conference on Nanomaterials and Advanced Composites*, Springer Proceedings in Physics 298,
https://doi.org/10.1007/978-981-99-7153-4_2

resins, which limits their range of industrial applications. For this reason, FRPs with thermosetting resin matrices are employed in many structural components that require high strength, such as aircraft parts. Epoxy resins are representative thermosetting resins that are widely used as FRP matrix materials because of their excellent water resistance, heat resistance, and adhesiveness to various surfaces. In these applications, FRPs are typically formed by laminating intermediate materials (called prepregs), which are fabricated by impregnating long fibers serving as a reinforcing material with an epoxy resin mixed with a hardener. Various prepreg types are available on the market, including those with fibers aligned in one direction and plain weave fibers impregnated with a resin.

Prepregs are widely utilized as intermediate products in the FRP fabrication process. Because a prepreg is impregnated with an epoxy resin mixed with a curing agent, a ring-opening polymerization of the resin occurs at room temperature, initiating a curing process. Therefore, FRP molding manufacturers must store prepregs in a freezer or other low-temperature facility after purchasing and process them as soon as possible. Table 2.1 lists various commercial prepregs and their storage lives. It shows that despite some variations in the curing temperature, a proper temperature control of the as-purchased prepregs represents a major challenge for molding manufacturers who desire to maintain the high product quality. If these prepregs could be stored at room temperature, the need for not only such management, but also bulky facilities such as freezers would be completely eliminated, which may help reduce the running costs and environmental burden by decreasing power consumption [1]. To the best of our knowledge, epoxy resins that can be stored semi-permanently at room temperature are not commercially available and have been evaluated only by Tran et al. [2] and Hayaty et al. [3]. Therefore, the mechanical properties of these resins have not been fully investigated, which limits the scope of their practical applications.

FRP prepregs containing epoxy resins often include bisphenol A and bisphenol F, and their curing procedure is performed at room temperature through an addition reaction with an amine compound. Therefore, such prepregs must be stored at low temperatures. However, epoxy resins can be originally cured by opening the epoxy ring and subjecting it to cationic polymerization in addition to the amine addition reaction. In this method, the ring-opening polymerization of epoxy rings is initiated by applying heat at a temperature of 100 °C or higher, and the curing process may be accelerated by adding a small amount of a curing agent acting as a catalyst to the epoxy resin [4, 5]. The resulting cured material is characterized by a high glass transition temperature. In the ring-opening polymerization process, the utilized heat treatment

Table 2.1 Shelf lives of TORAYCA prepregs [6]

	Storage temperature	Hardened at 130 °C	Hardened at 180 °C
Shelf life	−18 °C or less	6 months	6 months
	5 °C or less	3 months	2 months
	20 °C or less	1.5 months	14 days

Table 2.2 Physical properties of CF tow [7]

Type	Number of filaments	Filament diameter	Young's modulus	Tensile strength	Density
TR 50S12L	12,000	7 μm	235 GPa	4.90 GPa	1.82 g/cm^3

conditions are the main factors affecting the reaction rate and degree of curing. In this study, we first identified the optimum heating conditions for FRP matrices using epoxy resins mixed with a commercially available cationic polymerization initiator. After that, prepregs containing a room temperature-stable epoxy resin were prepared, and the mechanical properties of FRPs fabricated by laminating these prepregs were investigated to determine the applicability of room temperature-storable epoxy resins to FRP matrices.

Carbon fiber-reinforced plastics (CFRPs) are used in various fields (including the aerospace industry) because of their high specific strength and specific stiffness. If the prepregs utilized in the manufacture of such products can be stored at room temperature while maintaining their mechanical properties, the number of storage facilities and manufacturing costs may be significantly reduced, stimulating further industrial development. For this reason, carbon fibers were selected as the reinforcement material for FRP evaluation.

2.2 Materials and Methods

2.2.1 Fiber and Matrix Materials

Unidirectional carbon fiber bundles (PYROFIL TR 50S12L carbon fiber tow manufactured by Mitsubishi Chemical Corporation, hereinafter referred to as "CF tow") were used as the reinforcement material in this work. CF tow contains a bundle of 12,000 carbon fiber filaments with diameters of 7 μm taped together with a sizing agent. Table 2.2 lists the physical properties of CF tow. The matrix materials included jER828 (bisphenol A-type resin manufactured by Mitsubishi Chemical Corporation) and EP-4901 (bisphenol F-type resin manufactured by Adeka Corporation), which are general-purpose epoxy resins. Cationic polymerization initiator San-aid SI-100L (manufactured by Sanshin Chemical Industry, Co., Ltd.) was used as the curing agent. Table 2.3 lists the specifications of the utilized epoxy resins.

2.2.2 Neat Matrix Fabrication

SI-100L was added to 100 parts of the main agent, jER828 or EP-4901, to 2 parts of SI-100L, and mixed well. The obtained system was evacuated in a vacuum chamber

Table 2.3 Specifications of the epoxy resins used in this work [8, 9]

Epoxy resin	Epoxy equivalent	Viscosity, mPa·s/ 25 °C	Specific gravity at 25 °C	Appearance
jER828 (Bisphenol A-type)	184–194	12,000–15,000	1.17	Liquid, colorless, and transparent
EP-4901 (Bisphenol F-type)	170	3500	–	Liquid, colorless, and transparent

at an internal temperature of 30 °C until all air bubbles were eliminated. Afterward, the resin was poured into a Teflon mold with a rectangular shape having a length of 100-mm and width of 15 mm, sandwiched between two glass plates placed on the top and bottom, and cured in a constant temperature dryer in three stages. The first step was pre-curing at 80 °C for 2 h; the second step was post-curing at 120 °C for 1 h or 6 h; and the third (after-curing) step was conducted either at 150 °C for 2, 4, or 6 h, or at 180 °C for 2 h.

2.2.3 Prepreg and Composite Material Fabrication

The procedure described in Sect. 2.2 was used to prepare a de-aerated precursor by mixing an epoxy resin base with a hardener. The epoxy resin mixed with a hardener was applied inside a 100-mm square mold with a brush; two layers of CF tow were placed on the top of each other without distortion; the other side of the mold was also treated with a brush for resin impregnation. The obtained composite material was sandwiched in a hot press at a small applied pressure, pre-cured at 80 °C for 2 h, post-cured at 120 °C for 1 h, and after-cured at 180 °C for 4 h or 6 h.

2.2.4 Tensile Testing

After curing, the neat matrix was removed from the mold, and its four surfaces except for the end faces were polished with #180, #600, #1000, and #2500 emery paper to remove irregularities and finally buffed to a mirror finish. The composite specimens were cut to a width of 15 mm to ensure that CF tow fibers were oriented in the longitudinal direction, and the cut surfaces were polished with emery paper. Aluminum tabs with tapered edges were attached to the polished specimen to create a gripping portion. A schematic of the obtained specimen is shown in Fig. 2.1. The gauge length was 50 mm, and the tensile speed was 1 mm/min. A strain gauge was attached to the specimen center in the longitudinal direction to measure the strain and load to rupture.

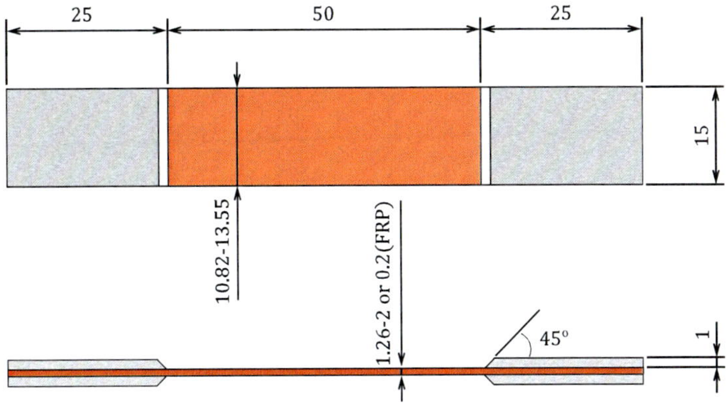

Fig. 2.1 Schematic of the tensile test specimen

2.2.5 Curing Degree Measurements

The reaction rates of the room temperature-storable epoxy resins were calculated by measuring the relative fractions of epoxy ring-openings in the neat matrix and composite specimens by attenuated total reflection Fourier–transform infrared (FT–IR) spectroscopy.

2.3 Results and Discussion

2.3.1 Neat Matrix Tensile Tests

2.3.1.1 Bisphenol A Epoxy Resin (jER828)

Table 2.4 lists the results of tensile tests and FT–IR curing degree measurements conducted for jER828 resin under various curing conditions. It shows that the reaction rate increases with increasing heating time and temperature. The highest tensile strength corresponds to the lowest reaction rate (No. 1), and its value decreases with the reaction rate. Specimen No. 1 exhibits the lowest Young's modulus; however, its elongation is more than twice that of the other specimens, indicating that the material exhibits the highest toughness. Because epoxy resins form a strong three-dimensional network structure with an opening of their rings, the reaction rate positively correlates with the tensile strength. When an ordinary polymerization-type curing agent is used, a blending amount of approximately 10–50% by weight is required, and when the same base epoxy resin is utilized, the cured material becomes hard because the amount of the added catalyst-type curing agent is larger than that used in the present study [10]. The obtained results revealed that the tensile strength decreased as the reaction rate increased, but Young's modulus values measured at the other conditions

Table 2.4 Results of tensile tests and reaction rate measurements of jER828/SI-100L

Condition No	Pre-cure	Post-cure	After-cure	Tensile strength [MPa]	Young's modulus [GPa]	Fracture strain [%]	Reaction rate [%]
1	80 °C 2 h	120 °C 1 h	–	59.6	1.86	3.23	57.8
2	80 °C 2 h	120 °C 6 h	–	48.5	3.09	1.62	75.6
3	80 °C 2 h	120 °C 1 h	150 °C 2 h	44.9	2.82	1.62	85.8
4	80 °C 2 h	120 °C 1 h	150 °C 4 h	40.2	2.61	1.60	89.3
5	80 °C 2 h	120 °C 1 h	150 °C 6 h	28.3	2.79	1.11	91.9
6	80 °C 2 h	120 °C 1 h	180 °C 2 h	26.0	2.72	1.16	95.4

exceeded that of specimen No. 1. Furthermore, the fracture strain decreased with increasing reaction rate, falling to the low 1% range at reaction rates above 90%, indicating brittle behavior due to hardening. When the resin is used as an FRP matrix, the heating conditions utilized for sample No. 1 are considered optimal because its fracture strain is higher than the fiber strain at break ($\varepsilon = 0.015$ for general carbon fibers). However, if an epoxy resin reaction rate of 90% or more is not achieved, the resin properties will change during the life cycle owing to the effects of ultraviolet rays and temperature.

2.3.1.2 Bisphenol F Epoxy Resin (EP-4901)

Bisphenol F epoxy resins are often used as FRP matrices because of their low viscosity and excellent moldability as compared with those of type A resins. Therefore, the room temperature applicability of these epoxy resins was investigated. Table 2.5 lists the test results obtained for each heating condition. Almost all EP-4901 specimens subjected to after-curing (Nos. 8–11) exhibited reaction rates of 90% or higher, indicating high property stability. In contrast to jER828, the tensile strength of EP-4901 did not decrease with increasing reaction rate, and its strength was at least 24% higher than that of jER828 with a maximum stress difference of 145%. Young's modulus of this resin remained above 3 GPa, even though it decreased by approximately 10% with increasing reaction rate. The fracture strain increased with increasing reaction rate, and no embrittlement was observed even when the reaction rate exceeded 90%. The maximum tensile strength was achieved under the No. 9 heating condition, and the strain at break exceeded 5%, suggesting that the tested material was considerably more ductile than the fibers. Based on these results, the applicability of the FRP prepreg and composite materials prepared under the heating conditions, at which the reaction rate exceeded 90% (Nos. 9–11 for EP-4901), was established.

Table 2.5 Results of tensile tests and reaction rate measurements of EP-4901/SI-100L

Condition No	Pre-cure	Post-cure	After-cure	Tensile strength [MPa]	Young's modulus [GPa]	Fracture strain [%]	Reaction rate [%]
7	80 °C 2 h	120 °C 1 h	–	60.4	3.58	2.88	68.1
8	80 °C 2 h	120 °C 1 h	150 °C 2 h	66.5	3.15	3.90	88.9
9	80 °C 2 h	120 °C 1 h	150 °C 6 h	76.7	3.19	5.34	90.7
10	80 °C 2 h	120 °C 1 h	170 °C 2 h	70.1	3.12	4.96	93.0
11	80 °C 2 h	120 °C 1 h	180 °C 2 h	75.8	3.29	4.98	91.7

2.3.1.3 Effect of Cycloaliphatic Epoxy Resin on Tensile Properties

Generally, when epoxy resins are synthesized, they are mixed with additives, such as antioxidants and flame retardants, according to their intended use and molded. For the bisphenol A-type epoxy resins examined in this study, as the reaction rate increased, the strength and fracture strain decreased, leading to material brittleness. Therefore, we attempted to prepare epoxy resins with flexibility and toughness that would maintain a high level of fracture strain even at large reaction rates. For this purpose, epoxidized polybutadiene (EPOLEAD PB3600, manufactured by Daicel Corporation) was added to each epoxy resin as an alicyclic agent, and the tensile properties of the resulting system were evaluated. In this experiment, 20% EPOLEAD was mixed with each epoxy resin base followed by the addition of 2% SI-100L. Table 2.6 lists the results of tensile tests conducted for the produced resins. It shows that the reaction rate decreases with the addition of EPOLEAD under all heating conditions. A comparison of the jER828 resins prepared with and without EPOLEAD (Nos. 2 and 12, Nos. 3 and 13, and Nos. 4 and 14) demonstrated that the tensile strength, Young's modulus, and fracture strain decreased after EPOLEAD addition. The decrease in Young's modulus was accompanied by softening, which is considered a significant effect of the lower reaction rate (compared to the no-mixing condition). The decrease in the fracture strain that occurred despite the decrease in the reaction rate suggests that the effect of EPOLEAD addition is not very strong and that EPOLEAD itself is susceptible to thermal degradation. A comparison of the EP-4901 resins prepared with and without EPOLEAD blending (Nos. 7 and 15, Nos. 9 and 16, and Nos. 11 and 17) revealed that all their tensile properties deteriorated after curing (Nos. 16 and 17) similar to jER828. However, under condition No. 15 without an after-curing stage, although the reaction rate was significantly lower than that obtained without EPOLEAD mixing, a higher specimen flexibility was achieved, as indicated by the decrease in Young's modulus and increase in the fracture strain, confirming the susceptibility of EPOLEAD to thermal degradation.

Figure 2.2 displays the stress–strain curves of specimens Nos. 1, 9, 13, and 15, which represent the best tensile properties of the bisphenol A and bisphenol F resins prepared with and without EPOLEAD addition. It shows that jER828 exhibits linear trends with and without EPOLEAD mixing, while EP-4901 demonstrates nonlinear

Table 2.6 Results of tensile tests and reaction rate measurements of both epoxy resins/SI-100L containing EPOLEAD

Condition no	Base epoxy	EPOLEAD content	Pre-cure	Post-cure	After-cure	Tensile strength [MPa]	Young's modulus [GPa]	Fracture strain [%]	Reaction rate [%]
12	jER828	20%	80 °C 2 h	120 °C 6 h	–	35.27	2.79	1.38	66.3
13			80 °C 2 h	120 °C 1 h	150 °C 2 h	38.13	2.48	1.60	72.1
14			80 °C 2 h	120 °C 1 h	150 °C 4 h	31.33	2.47	1.28	75.1
15	EP-4901		80 °C 2 h	120 °C 1 h	–	56.18	3.00	4.18	37.5
16			80 °C 2 h	120 °C 1 h	150 °C 6 h	43.65	2.88	2.10	81.0
17			80 °C 2 h	120 °C 1 h	180 °C 2 h	56.06	2.68	3.30	85.3

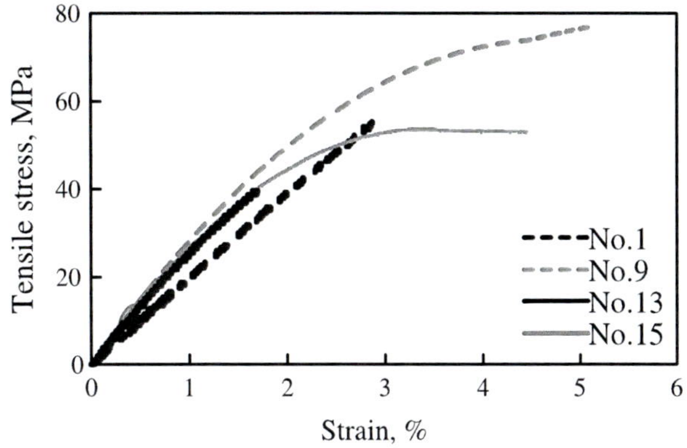

Fig. 2.2 Stress–strain curves recorded under the optimal heating condition for each epoxy resin

behavior starting from a strain of approximately 1–2%. The fracture strain of EP-4901, which is a bisphenol F-type resin, is higher than that of jER828 owing to its higher toughness. No embrittlement due to the larger reaction rate was observed, indicating that EP-4901 is more suitable for FRP matrices.

2.3.2 Composite Material Tensile Test

Table 2.7 lists the results of tensile tests of the CFRPs composed of the room temperature-stable epoxy resins, which were fabricated from EP-4901 without EPOLEAD and after-cured as described in the previous section. Because the CFRPs were heated by the upper and lower hot plates of the hot press, their reaction rates did not reach 90% during after-curing at 150 °C (unlike the neat matrix); therefore the after-curing process was conducted at 180 °C. The obtained results revealed that the fabricated CFRPs possessed high tensile strengths of 1265 MPa at a fiber volume fraction of 49.5% and 1582 MPa at a fiber volume fraction of 59.4%. Moreover, a 90% reaction rate was achieved after 2 h of after-curing at 180 °C, and Young's modulus exceeded 100 GPa, indicating high property stability. To investigate the storage stability of this epoxy resin at room temperature, EP-4901 was mixed with SI-100L and placed at room temperature for 1 week and 4 months followed by tensile testing. The obtained results are listed in Table 2.8. Specimens Nos. 18 and 20 exhibit little differences between their fiber volume fractions and tensile properties, indicating that their stability was not affected after 1 week of storage. Specimen No. 21 demonstrated superior tensile properties in spite of the lower fiber volume fraction than that of specimen No. 19. These results confirm that the studied material retained stable properties even after 4 months of storage, indicating its high performance as a room temperature-storable CFRP prepreg. Because the after-curing of specimen No. 19 was performed under the most severe conditions (at 180 °C for 6 h), the resin was likely overheated and partially subjected to thermal degradation. Therefore, for the bisphenol F-type resin EP-4901 used in this study, pre-curing was conducted at

80 °C for 2 h, post-curing—at 120 °C for 1 h, and after-curing—at 180 °C for 4 h using the hot press, which were considered the most suitable conditions for the FRP matrix.

In the next experiment, epoxy resin was manually applied to carbon fibers to create a pseudo-prepreg, which was then heat-cured inside the hot press. Because unidirectional materials were used in this study and tensile tests were conducted in the axial fiber direction, the stresses that fibers could sustain at break were calculated based on the law of mixture of stresses in the fiber direction expressed by Eq. (2.1).

$$\sigma_c = \sigma_f V_f + \sigma_m (1 - V_f) \tag{2.1}$$

$$\overline{\sigma_f} = \frac{\sigma_c^* - \sigma_{m'}(1 - V_f)}{V_f} \tag{2.2}$$

Here, Eq. (2.1) is transformed into the form represented by Eq. (2.2), where σ_c and σ_c^* are the tensile stress and tensile strength of the composite, respectively; V_f is the fiber volume fraction; σ_f is the fiber stress; σ_m is the matrix stress; $\overline{\sigma_f}$ is the stress that results in a composite rupture; $\sigma_{m'}$ is the matrix stress corresponding to the strain at the composite rupture. The $\sigma_{m'}$ value was obtained by assuming that the stress–strain curve recorded during the tensile test of the neat matrix of No. 9 had the same properties as those of the resin used for CFRP fabrication. The obtained results are presented in Fig. 2.3. It shows that the fibers were subjected to stresses greater

Table 2.7 Tensile test results of the CFRP specimens fabricated from the room temperature-stable epoxy resin

Condition no	Pre-cure	Post-cure	After-cure	Fiber volume fraction [%]	Tensile strength [MPa]	Young's modulus [GPa]	Fracture strain [%]	Reaction rate [%]
18	80 °C 2 h	120 °C 1 h	180 °C 4 h	49.5	1265	106.4	1.14	91.2
19	80 °C 2 h	120 °C 1 h	180 °C 6 h	59.4	1582	126.7	1.29	–

Table 2.8 Tensile test results of the CFRP specimens with the room temperature-stable epoxy resin obtained after different storage periods

Condition no	Storage period	Pre-cure	Post-cure	After-cure	Fiber volume fraction	Tensile strength [MPa]	Young's modulus [GPa]	Fracture strain [%]
20	1 week	80 °C 2 h	120 °C 1 h	180 °C 4 h	49.2	1264	107.1	1.2
21	4 months	80 °C 2 h	120 °C 1 h	180 °C 4 h	54.6	1743	130.4	1.45

than 2500 MPa under all conditions. In addition, specimen No. 21 was subjected to a stress of 3160 MPa. Thus, at least 65% of the maximum fiber stress of 4.9 GPa listed in Table 2.2 can be obtained by optimizing the heating conditions with the current resin composition (EP-4901: SI-100L = 100:2). Further performance improvement of carbon fibers could be achieved by minimizing the number of forming defects and increasing the matrix toughness. Typical stress–strain curves recorded for specimens Nos. 18–21 are shown in Fig. 2.4. Under all conditions, the stress exhibited an almost linear transition to rupture. During tensile testing, a fiber breaking sound was detected for the composite, and a slightly nonlinear part was also observed at a strain of approximately 0.9% for specimen No. 21 (Fig. 2.4), indicating that the room temperature-stable epoxy resin could sustain the load released by the breakage of carbon fibers, which confirmed the high composite toughness.

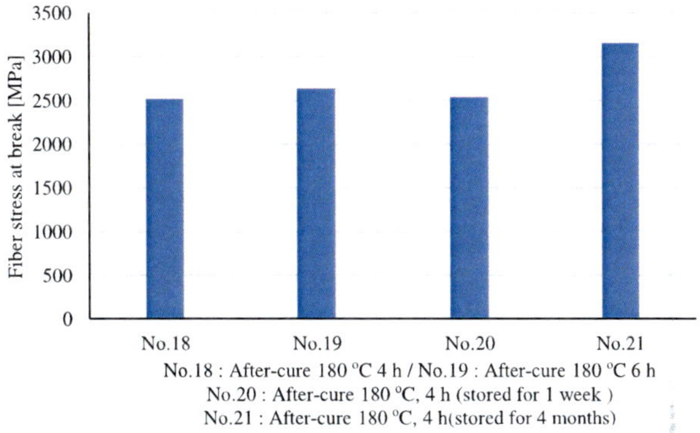

Fig. 2.3 Fiber stress at break determined under each testing condition

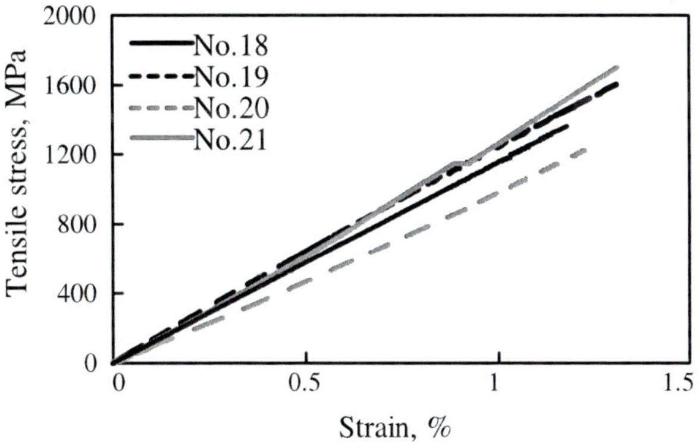

Fig. 2.4 Stress–strain curves of the CFRP specimens fabricated from the room temperature-stable prepreg

2.4 Conclusion

In this study, we investigated the tensile properties of the neat CFRP matrix and CFRP specimens using the epoxy resin prepared by the ring-opening polymerization of epoxy groups and catalytic curing agent. This method required heating during the curing process instead of adding a polymerization curing agent (such as an amine compound) to store the produced FRP prepreg at room temperature instead of low temperatures. The obtained results are summarized below.

(1) The ring-opening polymerization of epoxy rings progressed with increasing heating temperature and time for the samples containing bisphenol A jER828 resin and SI-100L. During this process, the reaction rate increased, but the tensile strength and fracture strain decreased significantly while Young's modulus increased, suggesting that the thermal degradation caused by heating could contribute to material embrittlement.
(2) In contrast to jER828 resin, the tensile strength of the specimens containing the bisphenol F-based EP-4901 resin and SI-100L increased with increasing reaction rate during heating, while Young's modulus did not change significantly and remained above 3.1 GPa. The maximum fracture strain exceeded 5% at a reaction rate of 90%, indicating that the resin possessed tensile properties applicable to FRP matrices.
(3) The addition of the alicyclic agent EPOLEAD with a flexible backbone did not improve the fracture strain for both types of epoxy resins. Furthermore, the tensile properties of the resins deteriorated and their reaction rates decreased under the conditions involving EPOLEAD addition. Because the material flexibility was improved by the introduction of EPOLEAD only for EP-4901 without an after-curing step, it was concluded that EPOLEAD was thermally degraded and embrittled by the heat treatment at temperatures exceeding 150 °C.
(4) The CFRP containing the EP-4901/SI-100L epoxy resin that was after-cured without adding EPOLEAD exhibited a reaction rate of over 90% and achieved high strength and Young's modulus by after-curing at 180 °C for 4 h before molding in the hot press.
(5) The calculated value of the maximum stress that could be sustained by composite material fibers was 3160 MPa. The maximum stress of the fibers was 4.9 GPa, and the fabricated CFRP could endure 65% of this value. The full fiber potential can be realized if the toughness of the room temperature-stable epoxy resin and its molding stability are further improved.
(6) The stress–strain curves of the fabricated CFRP samples revealed that the surrounding matrix endured the load released by the internal fiber breakage during tensile loading, suggesting that the EP-4901/SI-100L resin was highly suitable for FRP matrices.

References

1. Kataoka, Y., Nomura, T., Kawaoka, Y., Kouge, K., Goda, K.: Effect of epoxidized polybutadiene on tensile properties of pre-cured resin storable at room temperature. In: 11th Asian–Australasian Conference on Composite Materials, Abstract book distributed on USB, Cairns (2018)
2. Tran, A. D., Koch, T., Liska, R., Knaack, P.: Radical-induced cationic frontal polymerization for prepreg technology. Monatshefte für Chemie—Chemical Mon. 152, 151–165 (2021)
3. Hayaty, M., Honarkar, H., Beheshty, M.H.: Curing behavior of dicyandiamide/epoxy resin system using different accelerators. Iran. Polym. J. **22**, 591–598 (2013)
4. Onizuka, K.: Epoxy resin hardener. J. Adhes. Soc. Jpn. **53**(4), 122–128 (2017)
5. Koike, T.: Structure and properties of thermal cationic polymerization initiator as a latent curing agent for epoxy resin. J. Adhes. Soc. Jpn. **56**(1), 20–33 (2020)
6. Toray Industries, Inc. Homepage: https://www.cf-composites.toray/ja/resources/data_sheets/#anc3. last accessed 2022/10/20
7. Mitsubishi Chemical Corporation Homepage: https://www.m-chemical.co.jp/carbon-fiber/product/tow/. last accessed 2022/10/20
8. Mitsubishi Chemical Corporation Homepage: https://www.m-chemical.co.jp/products/departments/mcc/epoxy/product/1200246_7184.html. last accessed 2022/10/20
9. Adeka Corporation Homepage: https://www.adeka.co.jp/chemical/products/functional/pro142c.html. last accessed 2022/10/20
10. Tatsumi, A.: Latest development and application of cycloaliphatic epoxy resin. J. Adhes. Soc. Jpn. **53**(11), 391–397 (2017)

Chapter 3
Effect of Process Parameters on Feasibility of Production of Cellulose Nanofiber Yarn by Wet Spinning

Yoshinobu Shimamura, Hiroki Kondo, and Tomoyuki Fujii

Abstract Cellulose nanofibers (CNFs) have been attracting attention as engineering materials because CNFs have high tensile properties in addition to their biodegradability and reproducibility. Many research works have been conducted to produce high-performance composite materials using CNFs. Alignment of CNFs in composite materials is critical for achieving higher mechanical performance, but it is challenging because CNFs have high flexibility. Wet spinning of CNFs is a technique to obtain a pure CNF yarn by injecting CNF hydrogel into a liquid like acetone, followed by drying. The high shear rate in the injection nozzle gives the orientation of CNFs along with the flow direction, resulting in an almost unidirectional alignment of CNFs in the yarn. In this study, the effect of the inner diameter of the injection nozzle and the CNF concentration of hydrogel on the feasibility of the production of CNF yarns was experimentally investigated. The results showed that a limited range of the inner diameter of the injection nozzle enabled us to produce CNF yarns, but it was possible to produce CNF yarns within a certain range of CNF concentrations.

3.1 Introduction

Cellulose nanofibers (CNFs) have been attracting attention as engineering materials because they have high tensile properties in addition to their biodegradability and reproducibility. Young's modulus is thought to be 110–140 GPa, and the tensile strength is estimated to be 2–6 GPa [1]. Much research have been conducted to produce high-performance composite materials using CNFs. Alignment of CNFs in composite materials is critical for achieving higher mechanical performance, but it is

Y. Shimamura (✉) · T. Fujii
Shizuoka University, Hamamatsu 432-8561, Japan
e-mail: shimamura.yoshinobu@shizuoka.ac.jp

H. Kondo
Graduate School of Integrated Science and Technology, Shizuoka University, Hamamatsu 432-8561, Japan

R. Murakami et al. (eds.), *The 3rd International Conference on Nanomaterials and Advanced Composites*, Springer Proceedings in Physics 298,
https://doi.org/10.1007/978-981-99-7153-4_3

challenging because CNFs are highly flexible. Wet spinning of CNFs is a technique to obtain a pure CNF yarn by injecting CNF hydrogel into an organic solvent like acetone, followed by drying [2, 3]. The high shear rate in the injection nozzle gives the orientation of CNFs along with the flow direction, resulting in an almost unidirectional alignment of CNFs in the yarn. Thus, CNF yarns will be useful as a preform for fabricating unidirectionally aligned CNF-reinforced composites. Previous research on the wet spinning of CNFs has already shown the feasibility of wet spinning, but the process window has not been fully understood.

In this study, the effect of the CNF concentration of hydrogel and the inner diameter of the injection nozzle on the density of yarns was experimentally investigated.

3.2 Materials and Production of CNF Yarn

3.2.1 Materials

Hydrogel of 10 wt% cellulose nanofibers was purchased from Sugino Machine Ltd., Japan. Since the purchased hydrogel was too thick, the viscosity was decreased by diluting the hydrogel with purified water. Acetone (014-19095, Fujifilm Wako Pure Chemical Corporation, Japan) was used for solvent exchange after wet spinning.

3.2.2 Production of CNF Yarn

To produce CNF yarns, the diluted CNF hydrogel was injected into an acetone bath, forming a thread-like CNF gel as shown in Fig. 3.1. The thread-like CNF gel was kept in the acetone bath for 5 min for solvent exchange, and then dried for 24 h at room temperature. In this study, the flow rate was fixed to 45 mm/s but the inner diameter of the injection nozzle and the CNF concentration of hydrogel were changed. CNF hydrogels of 2.0–8.0 wt% were prepared and injection nozzles with the inner diameter of 0.41–1.20 mm were used for the experiments.

Typical SEM images of CNF yarns produced in our experiments are shown in Fig. 3.2. The CNF yarn was produced by using the injection nozzle with the inner diameter of 0.84 mm and the CNF concentration was 4.0 wt%.

3.3 Results and Discussion

Table 3.1 summarizes the experimental results to clarify the process window.

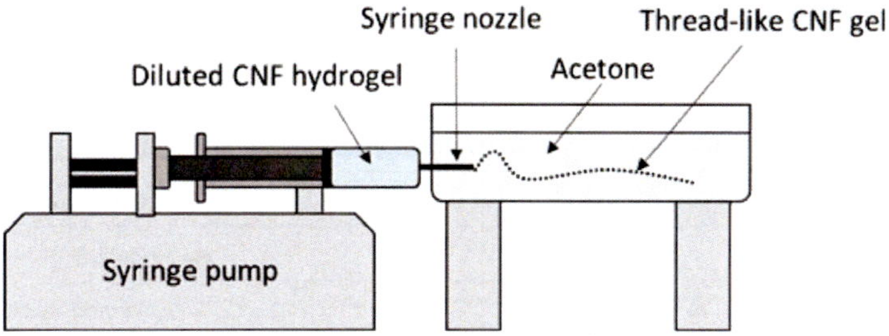

Fig. 3.1 Experimental set-up for wet spinning of CNF yarn

(a) Low magnification

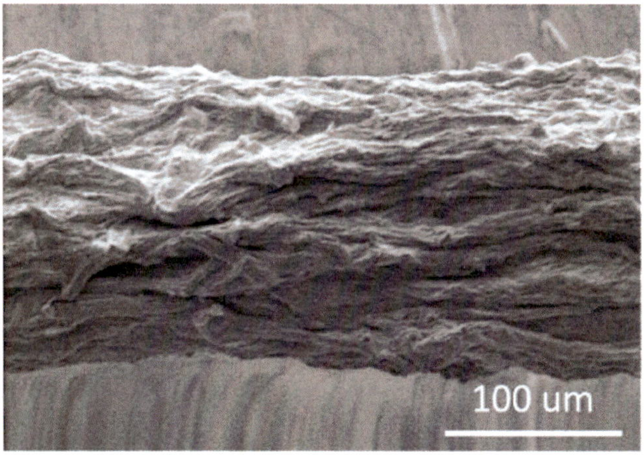

(b) High magnification

Fig. 3.2 SEM images of CNF yarn. **a** Low magnification, **b** High magnification

We found that CNF yarns could be produced in our case when the CNF concentration ranged from 3.0 wt% to 6.0 wt% and the inner diameter of the injection nozzle was around 0.9 mm. The experiments revealed that a limited range of the inner diameter of the injection nozzle enabled us to produce CNF yarns, but it was possible to produce CNF yarns within a certain range of CNF concentrations.

The relationship between the densities of CNF yarns and the inner diameter of the injection nozzle was plotted in Fig. 3.3, and the relationship between the densities of CNF yarns and the CNF concentration of hydrogel was plotted in Fig. 3.4.

Table 3.1 Summary of the feasibility of production of CNF yarn by wet spinning

CNF concentration (wt%)	Inner diameter of nozzle ϕ (mm)					
	0.41	0.60	0.70	0.84	1.04	1.20
2.0	×	×	×	×	×	
3.0			△	○		
4.0	▲	△	△	○	○	△
5.0				△	○	
6.0			△	△	○	▽
8.0				▼	▼	▽

○ : Possible
△ : Fragile when handling
▲ : Choking of nozzle
▽ : Injected gel unpickable
▼ : Almost impossible
× : Impossible

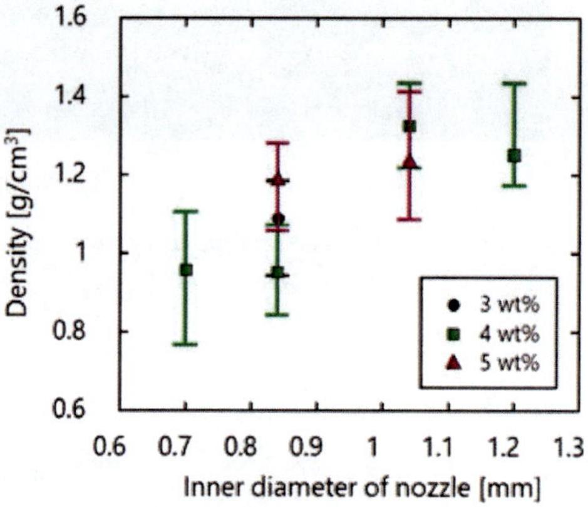

Fig. 3.3 Density of CNF yarns vs. Inner diameter of injection nozzle

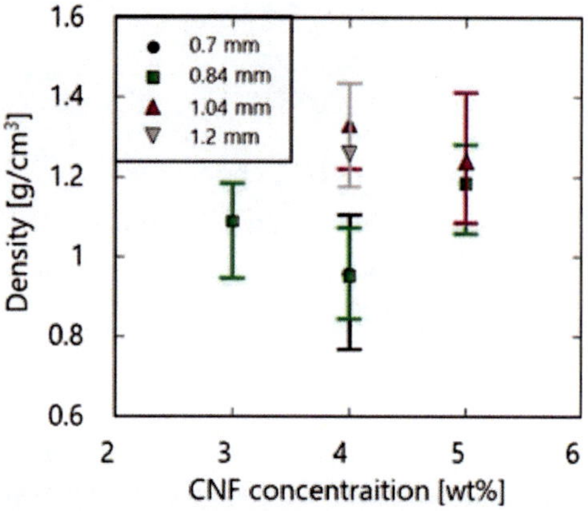

Fig. 3.4 Density of CNF yarns vs. CNF concentration of hydrogel

Roughly speaking, the density increased with increasing the inner diameter of the injection nozzle, but the dependency of CNF concentration was not obvious.

The density varied from 0.8 g/cm^3 to 1.4 g/cm^3, which means the porosity of a CNF yarn fell within 0.1–0.4 because the density of CNF is thought to be 1.5 g/cm^3 [4]. The porosity seemed to be appropriate for fabricating composite materials using CNF yarns as preform.

3.4 Conclusions

To clarify the effect of the inner diameter of the injection nozzle and the CNF concentration of hydrogel on the feasibility of CNF yarns, trials of the production of CNF yarns were conducted by changing the inner diameter of the injection nozzle and the CNF concentration of hydrogel. The results showed that a limited range of the inner diameter of the injection nozzle enabled us to produce CNF yarns, but it was possible to produce CNF yarns within a certain range of CNF concentrations of the hydrogel.

References

1. Saito, T., Kobayashi, Y., Fujisawa, S., Wu, C.-N., Isogai, A.: Fundamental Properties of Nanocellulose. Jpn. TAPPI J. **68**(8), 837–840 (2014)
2. Clemons, C.: Nanocellulose in spun continuous fibers: A review and future outlook. J. Renew. Mater. **4**(5), 327–339 (2016)
3. Lundahl, M.J., Klar, V., Wang, L., Ago, M., Rojas, O.J.: Spinning of cellulose nanofibrils into filaments: A review. Ind. Eng. Chem. Res. **56**(1), 8–19 (2017)
4. Sun, C.C.: Mechanism of moisture induced variations in true density and compaction properties of microcrystalline cellulose. Int. J. Pharm. **346**(1–2), 93–101 (2008)

Part II
Energy and Sensor Materials and Technology

Chapter 4
Synthesis of *N*-Methyl-D-Glucamine Modified Chitosan Nanofibers for Boron Adsorption

Masashi Kurashina, Daiki Kato, Haoyuan Li, Keita Shiba, Yuta Morishita, Kazuki Shibata, Ho Hong Quyen, and Mikito Yasuzawa

Abstract The *N*-methyl-D-glucamine modified chitosan nanofiber was synthesized using chitosan nanofiber. It was synthesized by reacting chitosan nanofiber with D-glucose and $NaBH_3CN$ to obtain D-glucamine-modified chitosan nanofiber, followed by the reaction with formic acid and formaldehyde. After swelling with water, the contained boron was removed by condensing with $NaHCO_3$. The maximum value of boron adsorption was 12.8 mg g^{-1} (equilibrium concentration, $C_e = 179$ ppm) and the value was higher than non-nanofiber *N*-methyl-D-glucamine modified chitosan obtained in the previous study.

4.1 Introduction

Boron compounds, such as boric acid, are widely used in glass, ceramic, detergent, and other industries. Since the high boron concentration is toxic to living organisms, effluent standards are set when boron is discharged into water bodies [1]. The amount of boron in drinking water according to WHO guidelines is 2.4 mg L^{-1} [2]. Amberlite IRA 743, which contains an *N*-methyl-D-glucamine functional group, is a well-known polymeric adsorbent for boron. This group is a type of ortho polyol, which consists of adjacent hydroxyl groups that adsorb boric acid and form an ester [3]. The problem is that the adsorbents cannot be naturally decomposed because they are mainly synthetic polymers. Therefore, it is necessary to develop a new adsorbent that is economical and environmentally friendly. Chitosan is a kind of natural

M. Kurashina (✉) · D. Kato · H. Li · K. Shiba · Y. Morishita · K. Shibata · M. Yasuzawa
Department of Applied Chemistry, Graduate School of Science and Technology, Tokushima University, 2-1 Minamijosanjima-cho, Tokushima-shi, Tokushima 770-8506, Japan
e-mail: kurashina.masashi@tokushima-u.ac.jp

H. H. Quyen
Faculty of Environment, University of Science and Technology, The University of Da Nang, 54 Nguyen Luong Bang, Da Nang, Vietnam

R. Murakami et al. (eds.), *The 3rd International Conference on Nanomaterials and Advanced Composites*, Springer Proceedings in Physics 298,
https://doi.org/10.1007/978-981-99-7153-4_4

polymer, easy to be chemically modified, environment-friendly, and biodegradable. Our previous reports showed the syntheses of glycosylated [4] and *N*-methyl-D-glucamine modified [5] chitosan-based boron adsorbents and their boron adsorption capacity. The maximum boron adsorption per weight of adsorbent was 5.8 mg g^{-1} for the former and 7.3 mg g^{-1} for the latter. The adsorption isotherms of the former followed the Langmuir model well, while the latter lacked reproducibility in the amount of adsorption. Additionally, chitosan nanofiber has been manufactured as a dispersed solution, and its specific surface area is 100 times higher than that of general chitosan flake. In this study, we synthesized *N*-methyl-D-glucamine modified chitosan from chitosan nanofibers to increase the amount of boron adsorption and measured its boron adsorption capacity.

4.2 Experiments

The synthesis of D-glucamine and *N*-methyl-D-glucamine modified chitosan nanofiber is depicted in Fig. 4.1. To synthesize the D-glucamine modified chitosan nanofiber, the procedure outlined in the literature [5, 6] was followed. Initially, 192 g of chitosan nanofiber (2 wt% from Sugino Machine Ltd.), 1.2 g of acetic acid, and 120 mL of methanol were mixed and stirred until the chitosan nanofiber was dissolved. Then, a mixture of D-glucose (12.9 g) and $NaBH_3CN$ (10.2 g) in 86 mL of water was slowly added to the chitosan nanofiber solution while stirring. The resulting soft white solid was separated via vacuum filtration after being stirred for 24 h and washed with methanol to yield the D-glucamine-modified chitosan nanofiber.

The *N*-methyl-D-glucamine modified chitosan nanofiber was synthesized in accordance with the procedure outlined in the literature [5, 7]. An amount of 14.6 g of D-glucamine modified chitosan nanofiber was mixed with 26.3 mL of formic

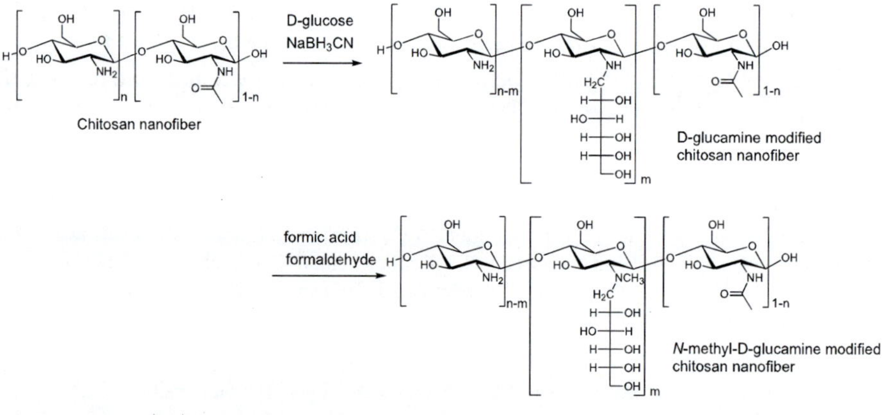

Fig. 4.1 Synthesis scheme of the D-glucamine and *N*-methyl-D-glucamine modified chitosan nanofiber

acid, 35.6 mL of formaldehyde, and 158 mL of water. The mixture was heated to 70 °C and kept at that temperature for 118 h. The resulting solution was evaporated to reduce the water content and then dropped into a 3 mol L^{-1} NaOH solution. The *N*-methyl-D-glucamine-modified chitosan nanofibers were obtained in the form of beads.

To test the effectiveness of boron adsorption, solutions of $Na_2B_4O_7 \cdot 10H_2O$ were prepared at concentrations of 200 ppm and 400 ppm (boron concentration). These boron solutions contained $NaHCO_3$ equivalent to 0.5 mol L^{-1}. *N*-methyl-D-glucamine modified chitosan nanofibers were then added to 20 mL of the solutions with varying concentrations of boric acid, with the initial concentration denoted as C_i [ppm as B]. The mixture was agitated in an isothermal bath at 25 °C for 24 h. The equilibrium concentration of the filtrate was noted as C_e [ppm as B], and both C_i and C_e were determined using the azomethine H absorptiometric method [8]. The amount of boron adsorbed was denoted as W [mg g^{-1}] and calculated as $W = (C_i - C_e) \times$ volume of water [L]/weight of adsorbent [g].

4.3 Results and Discussions

The yielded *N*-methyl-D-glucamine modified chitosan nanofiber contained boron because $NaBH_3CN$ was used in the reaction to modify the chitosan nanofiber with D-glucamine. The boron concentration increased when the *N*-methyl-D-glucamine modified chitosan nanofiber was used as is for boron adsorption experiments because boron was released without being adsorbed. The Boron desorption in acidic aqueous solutions was impossible because this material is readily soluble in acids. There is literature [9] on the insolubilization of acid-soluble chitosan by cross-linking with EGDE (ethylene glycol diglycidyl ether), but this material did not react with EGDE. This material could not even be washed with water because it swells in water and becomes a fragile gel. Figure 4.2(a) shows the loss of liquid phase as the material swells as a gel. Since this material precipitates in basicity, the boron was removed by condensing it once swollen with water and then adding $NaHCO_3$ to bring the solution to 0.5 mol L^{-1}. Figure 4.2(b) shows the material precipitating as an irregularly shaped gel. This process of precipitation with $NaHCO_3$ after swelling with water was repeated six times until no boron was detected in the filtrate.

Figure 4.3 shows C_e, equilibrium concentration, *vs.* W, the adsorbed amount of boron. The amount of boron adsorbed by the *N*-methyl-D-glucamine modified chitosan was $W = 8.6$ and 12.8 mg g^{-1} for $C_e = 56.7$ and 179 ppm, respectively. While the conditions for boron solutions were not the same, these values were greater than the boron adsorption of Amberlite IRA 743 and non-nanofiber *N*-methyl-D-glucamine modified chitosan obtained in the previous study [5]. This result was caused by the large surface area of the nanofibered chitosan, which resulted in a large amount of modification to the *N*-methyl-D-glucamine group, and efficient contact with the aqueous solution during boron adsorption.

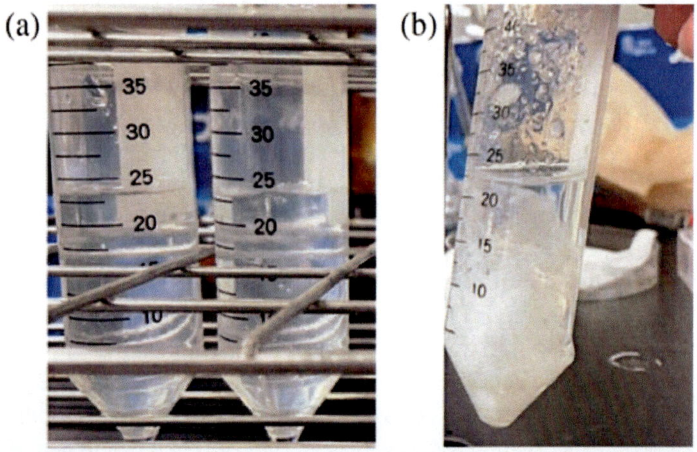

Fig. 4.2 The *N*-methyl-D-glucamine modified chitosan nanofiber was swollen with water (**a**) and precipitated with $NaHCO_3$ (**b**)

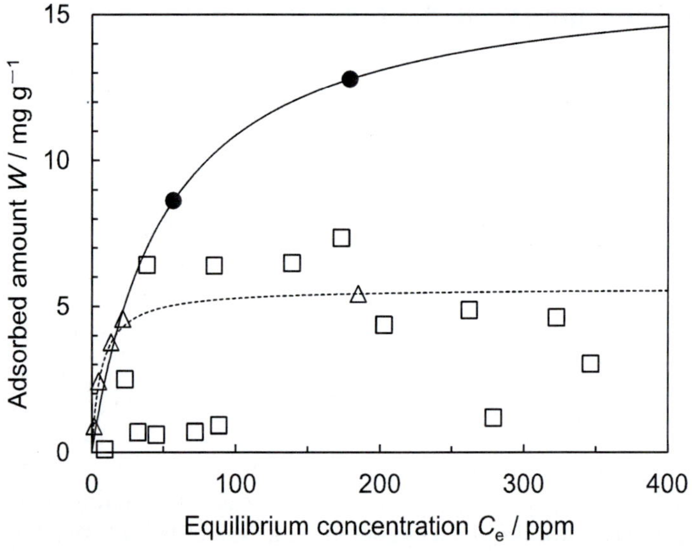

Fig. 4.3 Adsorption of boron using *N*-methyl-D-glucamine modified chitosan nanofiber (filled circle). The solid line follows the Langmuir model. Triangles and squares indicate the amount of boron adsorbed by Amberlite IRA 743 (dotted line follows Langmuir model) and *N*-methyl-D-glucamine modified chitosan from the previous study [5], respectively

4.4 Conclusions

The *N*-methyl-D-glucamine modified chitosan nanofiber was synthesized using chitosan nanofibers. The maximum value of boron adsorption was 12.8 mg g^{-1} (C_e = 179 ppm) and the value was higher than non-nanofiber *N*-methyl-D-glucamine modified chitosan obtained in the previous study [5]. Using nanofibered chitosan as a base material can yield more highly functional adsorbent.

References

1. Wang, B., Guo, X., P. Bai, P.: Removal technology of boron dissolved in aqueous solutions – A review. Colloids Surf. A: Phys.Chemical Eng. Asp. **444**, 338–344 (2014)
2. World Health Organization: Boron in drinking-water: background document for development of WHO guidelines for drinking-water quality. World Health Organization, Geneva (2009)
3. Wada, Y., Matsukami, T., Mori, S.: Ultra-efficient boron removal technology from water. KAGAKU SOUCHI (Plant and Process) **54**, 53–58 (2012). (in Japanese)
4. Ho Hong, Q.: Synthesis of eco-friendly adsorbents for the removal of contaminants in wastewater, Doctoral thesis, Tokushima University (2019). https://repo.lib.tokushima-u.ac.jp/113385, last accessed 2022/11/30
5. Kurashina, M., Li, H., Shiba, K., Morishita, Y., Shibata, K., Yasuzawa, M., Ho Hong, Q.: Syntheses of D-glucamine and *N*-methyl-D-glucamine modified chitosan for boron adsorption. Mod. Phys. Lett. B **36**(16), 2242001 (2022)
6. Yalpani, M., Hall, L.D.: Some chemical and analytical aspects of polysaccharide modifications. 3. Formation of branched-chain, soluble chitosan derivatives. Macromolecules **17**, 272–281 (1984)
7. Verheul, R.J., Amidi, M., van der Wal, S., van Riet, E., Jiskoot, W., Hennink, W.E.: Synthesis, characterization and in vitro biological properties of *O*-methyl free *N*,*N*,*N*-trimethylated chitosan. Biomaterials **29**(27), 3642–3649 (2008)
8. Ogawa, T.: Spectrophotometric determination of boron in glass with 1-(salicylideneamino)-8-hydroxynaphthalene-3,6-disulfonic acid. Bunseki Kagaku **35**(8), 709–712 (1986)
9. Wan Ngah, W.S., Ab Ghani, S., Kamari, A.: Adsorption behaviour of Fe(II) and Fe(III) ions in aqueous solution on chitosan and cross-linked chitosan beads. Biores. Technol. **96**(4), 443–450 (2005)

Chapter 5
Practical Microfluidic Technologies for In-Vitro Diagnostic Devices

Yusuke Fuchiwaki

Abstract This article reports the following three types of practical microfluidics: a technology for printing proteins such as antibodies on the surface of microfluidic channels, an ultra-rapid flow-through PCR technology, and a microchannel technology for rapid antigen–antibody reactions. The first is a microfluidic channel based on laser modification and printing technology. The surface of the acrylic substrate of the microfluidic channel was chemically modified by femtosecond laser ablation to facilitate the binding of antibodies. Antibodies can be printed on the acrylic substrates by dispensing an antibody solution with an inkjet printer. The second is the ultra-rapid flow-through PCR microfluidics device. This liquid-plug-flow through PCR effectively avoids flow instability caused by air bubbles and achieved 67% amplification compared to a thermal cycler in the 280 s for PCR completion. The third has developed a microfluidic device that allows enzyme-linked immunosorbent assay to be completed within 15 min by simply dropping specimens or reagents. These microfluidic technologies provide practical elemental technologies toward industrial applications.

5.1 Introduction

A point-of-care testing (POCT) for in vitro diagnostic (IVD) has become a major trend as a method of testing at or near the patient's bedside. POCT can provide results significantly faster than testing in a central clinical laboratory in a large hospital. Medical fields have emphasized the need for accurate, faster, low-cost diagnostic

Y. Fuchiwaki (✉)
Health and Medical Research Institute, AIST, Hayashi-cho, Takamatsu 2217-14, Kagawa, Japan
e-mail: yu-fuchiwaki@aist.go.jp

New-Generation Medical Treatment and Diagnosis Research Laboratory, AIST, 1-1-1 Higashi, Tsukuba, Ibaraki, Japan

Advanced Photonics and Biosensing Open Innovation Laboratory, AIST-Osaka Unverisity, Photonics Center, Osaka University, 2-1 Yamadaoka, Suita, Osaka, Japan

R. Murakami et al. (eds.), *The 3rd International Conference on Nanomaterials and Advanced Composites*, Springer Proceedings in Physics 298,
https://doi.org/10.1007/978-981-99-7153-4_5

devices. During the SARS-CoV-2 pandemic, fast testing kits for easy-of-use were indispensable tools. Thus, POCT is increasingly important in remote and low-income settings.

The in vitro diagnostics market has a growing need for high-throughput screening methods for the analyses of small sample volumes. Microfluidic devices, in which reagents and samples are used to perform reactions in a micro-space, are known as a new technology that has the potential to meet these challenges. As the continues moving to POCT, microfluidics devices are the ideal way, and many diagnostics companies are currently innovating in this field. For the practical application, we see accurate and stable analyses through simple operation as being critical. Thus, we focused on three developments of the microfluidics: a technology for printing proteins such as antibodies on the surface of microchannels, a practical high-speed PCR technology, and a microchannel device for the rapid antigen–antibody reactions.

The immunoassay microfluidics technology using enzyme-linked immunosorbent assay (ELISA) offers the accurate and sensitive determination of the analyte, but conventional ELISA requires large amounts of reagents and time-consuming incubation and washing processes, making it unsuitable for POCT. To solve this, much effort has been devoted to realizing the microfluidic-based immunoassay methods. Among them, fluidic control and antibody immobilization are crucial for the practical immunoassay microdevices. We implemented antibody immobilization on the surface of a plastic microchannel for ELISA using a piezoelectric inkjet printing system. Before the inkjet deposition of the antibody solution, the surface of the microchannel was ablated by UV-laser irradiation. UV photo-degradation reactions of polymethyl methacrylate (PMMA) involve main chain cleavage, incomplete or complete side chain cleavage, and direct UV-induced depolymerization. We performed the deposition of the first antibody on a UV-ablated PMMA microchannel surface. Carboxy-terminal propeptide of type I procollagen (PICP), a biomarker for osteoporosis and osteoblastic bone metastasis in prostate cancer, was employed as a model specimen to examine the performance of antibody immobilization by ELISA.

Secondly, we developed the ultra-rapid flow-through PCR microdevice. Generally, flow-through PCR device requires technical experience, skill, and knowledge, whereas we optimized the various conditions of the chip device for successful amplification by the liquid-plug-flow through PCR. So, this manuscript shows that our flow-through system is excellent in practical application.

Thirdly, I report a practical application of high-performance lateral-flow immunoassay based on the autonomous replacement of microfluidics. This microfluidic chip device is a novel lateral-flow test chip, which can perform the operation of the pump by just depositing the solution. This paper describes these approaches.

5.2 Capillary Driven-Flow Microfluidics

The microchannel was packaged using a film (polymethylmethacrylate, 0.047 mm in thickness, and 0.033 mm-layer adhesive, Toyo Ink Mfg. Co., Ltd., Tokyo, Japan). Reagents and samples were deposited at the inlet port, and the microchannels were filled with capillary force (Fig. 5.1a and b). The microchannels were vacated by the capillary action of paper in touch with the outlet between the immunoassay step (Fig. 5.1a).

Antibody immobilization was performed in two steps: UV-laser ablation of the PMMA microchannel surface and antibody inkjet printing. The laser ablation of the PMMA surface enabled both precise and digital control, and inkjet printing allowed high-throughput antibody patterning.

Inkjet printing (Cluster Technology Corporation, Osaka, Japan) was performed using a piezoelectric-driven drop-on-demand type. The inkjet system printed the primary antibodies in the spotting buffer on the microchannel surface. The PulseInjector® allowed easy adjustment of pico-liter droplets ejected. A differential drive

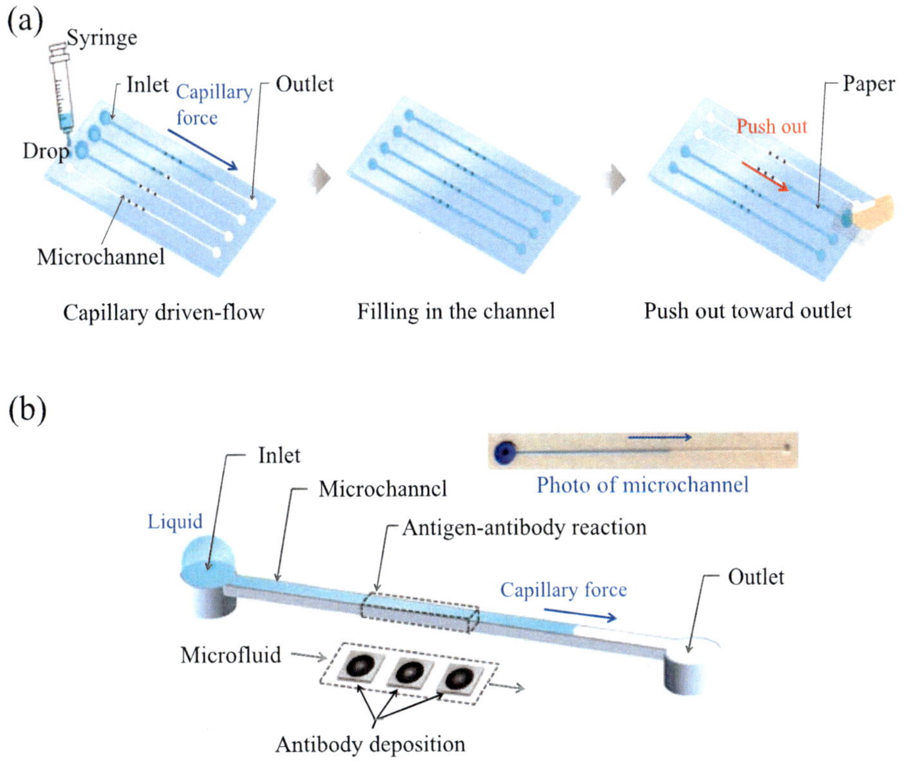

Fig. 5.1 Capillary-driven microchannel and UV-laser-based antibody immobilization procedures: **a** capillary-driven flow-based immunoassay process in operation; **b** single microchannel configuration during capillary-driven flow [1]

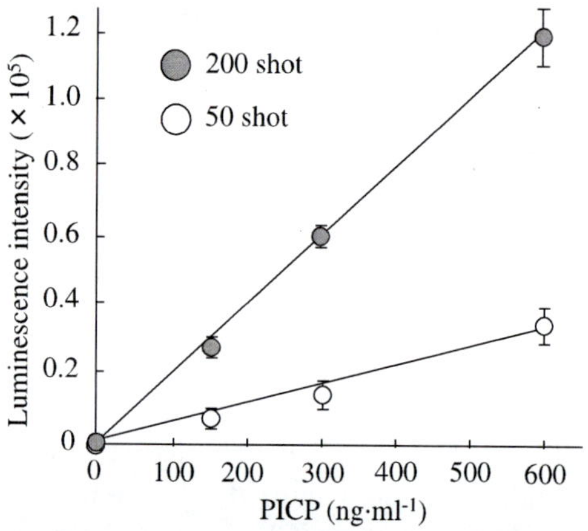

Fig. 5.2 Standard curve of 50 shots and 200 shots of antibody solution dispensed using a piezoelectric inkjet printing. Intensity increased linearly in a PICP concentration-dependent manner

waveform ensured stable ejection, and droplets were ejected at a frequency of 1 kHz and injection voltage of 8 V. The volume of one droplet of the primary antibody was 50 pL. Fifty and two hundred shots were printed onto the microchannel surface, comparing the luminescence intensities by the ELISA (Fig. 5.2).

Signal intensity was compared between 50 and 200 shots. The sensitivity of 200 shots was higher than that of 50 shots. This indicates that antibodies were immobilized at high density on the laser-ablated surface. The sensitivity and reproducibility demanded by sandwich ELISA require high-density immobilization of primary antibodies on the surface of the microchannel. Thus, the antibody immobilization technique of this study was effective in increasing the sensitivity of the microfluidic immunoassay chip.

5.3 Ultra-Rapid Flow-Through PCR Technology

The polymerase chain reaction (PCR) is a promising technique for the diagnosis of infectious diseases because it amplifies DNA specifically. PCR uses a three-step temperature cycle to make copies of specific DNA fragments, but each cycle takes about three minutes, so the complete reaction takes two to three hours. To overcome this issue, we developed the design of practical flow-through PCR microfluidic devices. Flow-through PCR microfluidic systems that enable fast, small-volume DNA amplification on a single chip are significantly impacting medical and bioanalytical research.

Flow-through PCR consists of a system with multiple temperature zones and long microfluidic channels to carry the PCR reagents to the respective temperature areas, allowing for fast DNA amplification without being affected by the heat transfer coefficient of the chamber material (Fig. 5.3) [2].

However, conventional flow-through PCR systems have problems such as requiring external devices for flow control, large amounts of reagents, and the need to

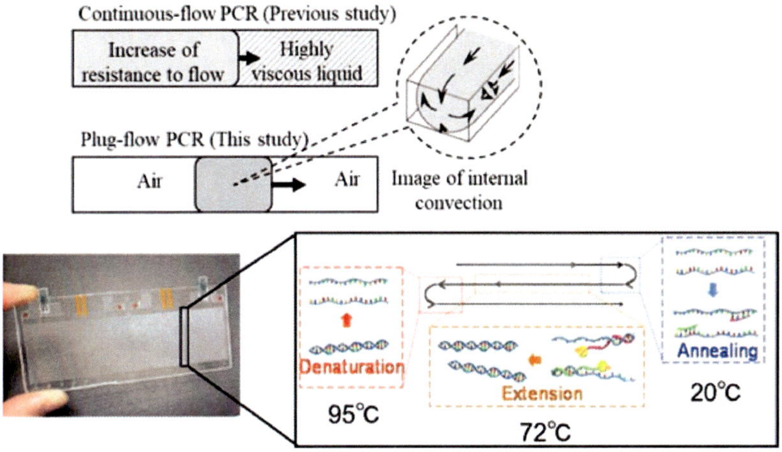

Fig. 5.3 Liquid plug flow-through DNA amplification in microchannel

avoid flow instability due to the generation of bubbles in the denaturation area. Therefore, we focused on the use of the thermal gradient generated in the microchannel as an actuator to implement plug-flow amplification.

Yields of PCR amplification between positive control and negative control were investigated against cycling time (Fig. 5.4) [3]. This value was then normalized to the product fluorescence from a standard thermal cycler as a baseline and plotted against total cycle time. The results showed 67% fluorescence at a flow time of 280 s, compared to less than 10% for the negative control. Amplification products were also detected with sufficient sensitivity. Fluorescence increased as a sigmoid function, which was a typical curve for PCR.

The fluorescent intensities obtained with plug-flow chips were lower than those observed with conventional thermal cyclers. This is due to the long microchannels, which have a large surface area to volume ratio, causing PCR components to adsorb on the chip surface. There is a risk of decreased amplification efficiency in systems using

Fig. 5.4 PCR amplification yield versus thermal cycler and cycling time [3]

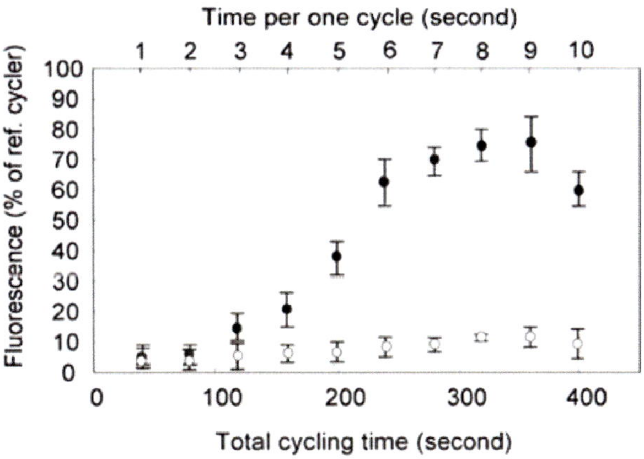

microchannels. The flow-through thermal cycling for the PCR has been performed by a continuous flow method, leading to complex operations requiring large amounts of the PCR reagents.

On the other hand, this plug-flow PCR system uses only the suction of the plug liquid by the pumping force. Therefore, this PCR system will have a significant impact on clinical applications toward on-site rapid diagnostics [4].

5.4 Accurate and Rapid Microfluidic ELISA System

In this study, we developed a new lateral-flow test chip that can perform pump-like actions simply by applying the solution to the inlet: flowing, stopping, and replacing the solution. The chip was prepared by bonding paper, film, and adhesive tape together. For sensitivity and accuracy in immunoassay detection, transparency and flatness of the substrate are important to analyze the weak light generated by a particular antigen–antibody response, but the paper is not flat and is opaque. So, capillary-driven-flow microfluidics are well-known to researchers.

In principle, aqueous solutions of samples and cleaning solutions should move without resistance from capillary forces, since solutions flowing by capillary forces have a strong tendency to stay in the microchannel. Even though a second drop of solution is added at the inlet, no solution replacement occurs.

Therefore, the solution flow in the developed device is driven by lateral flow, not capillary forces. The mechanism of solution replacement is shown in Fig. 5.2. The drop of solution at the inlet flows into the microchannel, and the excess is absorbed by absorbent paper located at the downstream of the microchannel (Fig. 5.2a). The excess solution is absorbed by the absorbent paper, leaving a solution in the microchannel that is equal to the microchannel volume (Fig. 5.2b).

The antibody captures antigen into each microchannel, and the solutions for sandwich immunoassay flowed into the microchannel, allowing luminescence intensity by the respective assays.

As a result, the novel chip device showed a good correlation in the range of 0.1–100 ng/ml of CRP concentration. The intensity of the chip device was higher than that of the cellulose membrane. Also, the cellulose membrane chip had a lot of nonspecific adsorptions compared to this chip device. The limitation of detection (LOD) of our chip device was 0.1 ng/ml (Table 5.1), whereas the LOD of the cellulose membrane was 100 ng/ml (Fig. 5.5).

There are no reports demonstrating that such sensitive detection is achieved by simply allowing the sample and reagent to deposit on the inlet. This chip system would show a significant impact in the fields of clinical point-of-care testing [6, 7].

Table 5.1 Immunoassay performance of stop & flow chip compared to nitrocellulose membrane [5]

	This study	Nitrocellulose membrane
LOD	<0.1 ng/mL	100 ng/mL
Sample volume	3–10 μL	3–10 μL
Nonspecific adsorption	Little or none	High

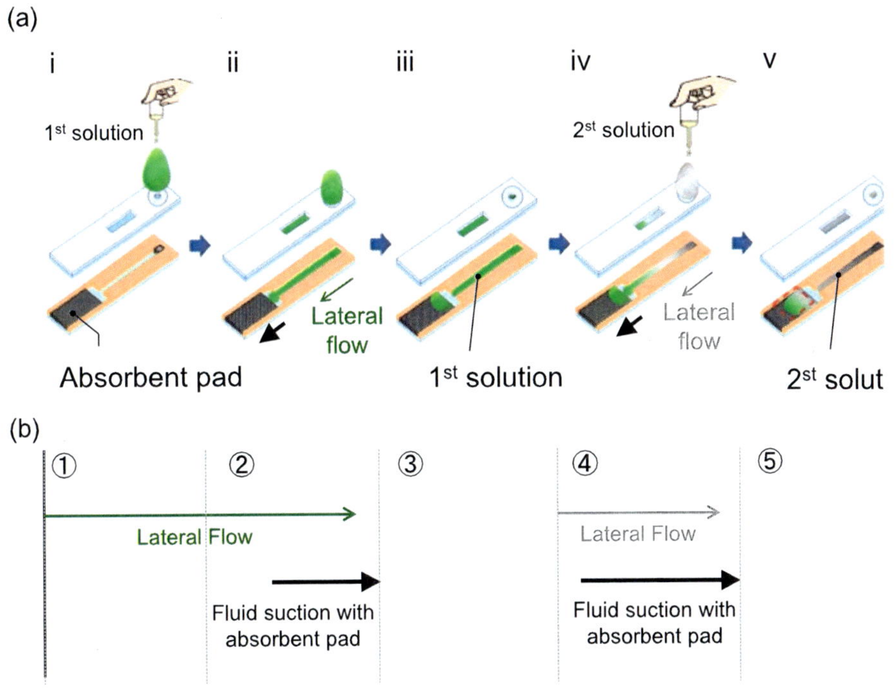

Fig. 5.5 Operation principle of solution exchange. **a** Diagram of the microfluidic flow during solution exchange. **b** Multiple driving forces continuously generated during solution exchange [5]

5.5 Conclusion

UV nanosecond laser ablation of PMMA has been shown to enable efficient and general-purpose antibody immobilization. Quantitative analyses of PICP concentrations yielded linear relationships in the following ranges: 0–600 ng ml^{-1} PICP. Antibody immobilization after UV-laser ablation is an excellent assay process that allows for precise manipulation and flexibility in mechatronic fabrication.

The developed plug-flow through PCR amplified about 67% of the samples compared to the thermal cycler of a commercial PCR device, and the time required was only 280 s. This liquid-plug-flow through PCR system was more practical than other systems.

The new lateral-flow test chip accomplished the operations like pump, such as flowing, stopping, and replacing solution, simply by dropping onto a port. The new chip showed a good correlation in the CRP concentration range of 0.1–100 ng/ml.

Acknowledgements Funding from the AMADA Foundation (AF-2020219-B3) and a Grant-in-Aid for Scientific Research (21K03798) from the Japan Society for the Promotion of Science is gratefully acknowledged. I am grateful to Dr. Masato Tanaka, Dr. Toshihiko Ooie, and Dr. Hidenori Nagai for their helpful discussions.

References

1. Fuchiwaki, Y., Takaoka, H.: UV-laser-assisted modification of poly (methyl methacrylate) and its application to capillary-driven-flow immunoassay. J. Micromech. Microeng. **25**, 075008 (2015)
2. Fuchiwaki, Y., Saito, M., Wakida, S.-I., Tamiya, E., Nagai, H.: A practical liquid plug flow-through polymerase chain-reaction system based on a heat-resistant resin chip. Anal. Sci. **27**, 225–230 (2011)
3. Fuchiwaki, Y., Nagai, H., Saito, M., Tamiya, E.: Ultra-rapid Flow-through polymerase chain reaction microfluidics using vapor pressure. Biosensensors Bioelectron. **27**(1), 88–94 (2011)
4. Nagai, H., Fuchiwaki, Y.: US Patent 10273532 (2019/04/30)
5. Fuchiwaki, Y., Goya, K., Tanaka, M.: Practical high-performance lateral flow assay based autonomous microfluidic replacement on a film. Anal. Sci. **34**, 57–63 (2018)
6. Youtube JP: https://www.youtube.com/watch?v=aD547atEDJA (2022/08)
7. Fuchiwaki, Y., Ooie, T., Kataoka, M., Takaoka, H.: US Patent 9744534 (2017/08/29)

Part III
Advanced Nanomaterials and Nanotechnology

Chapter 6
In-Situ Growth of Silicon Nanowires Array and Its Field Emission Behavior

Check for updates

Amol B. Deore, Krishna Jagtap, Sandesh R. Jadkar, and Mahendra A. More

Abstract *In-situ* growth of silicon nanowires (Si-NWs) array on Si substrate has been obtained by a simple approach based on physical masking. A physical mask comprised of an array of laser drilled circular holes (diameter ~20 μm) onto a thin tungsten foil was used to pattern Tin (Sn) thin film, used as a catalyst. The Si-NWs were selectively grown on the patterned Sn islands using the hot filament chemical vapor deposition (HFCVD) method under optimized experimental conditions. The as-synthesized array of Si-NWs was characterized using field emission scanning electron microscope (FESEM), UV–visible, and Raman spectroscopy. The field emission (FE) characteristics of the Si-NWs array emitter were investigated at a base pressure ~1 × 10^{-8} mbar. The emitter showed turn-on and threshold field values of 1.8 and 2.2 V/μm, respectively, which are lower than those reported for un-patterned Si-NWs and similar emitters. Interestingly, the maximum emission current density of ~1058.63 μA/cm^2 was extracted at a relatively lower applied field of 3.9 V/μm. Furthermore, the Si-NWs array emitter exhibited excellent emission stability at pre-set value ~1 μA over three hours duration. The results obtained herein clearly demonstrate the potential of the Si-NWs array emitter for practical applications in various vacuum micro-nanoelectronic devices.

6.1 Introduction

Field emitters based on one-dimensional (1D) nanostructures offer unprecedented advantages such as relaxation in operating pressure and voltages and have been realized for applications in various nano-microelectronic devices. Among the variety of 1D nanostructures, Si-NWs are attractive because of the easily understood electronic properties, quantum confinement effect, and ubiquitous role in electronic devices [1, 2]. The Si-NWs have been extensively studied for their applications in batteries,

A. B. Deore · K. Jagtap · S. R. Jadkar · M. A. More (✉)
Centre for Advanced Studies in Materials Science and Condensed Matter Physics, Department of Physics, Savitribai Phule Pune University, Pune 411007, India
e-mail: mam@physics.unipune.ac.in

R. Murakami et al. (eds.), *The 3rd International Conference on Nanomaterials and Advanced Composites*, Springer Proceedings in Physics 298,
https://doi.org/10.1007/978-981-99-7153-4_6

sensors, and solar cells [3–9], as well as field emission (FE)-based applications. Owing to inimitable geometrical shape and size, and excellent mechanical strength, the Si-NWs are recognized as an optimistic field emitter for X-ray generators [10, 11]. The FE characteristics can be improved further via synthesizing emitters comprised of an array of vertically oriented 1D nanostructures. For the synthesis of array-based structures, the use of lithography techniques is inevitable. The lithography techniques require special chemicals and control the environment, making these techniques complex and expensive. Therefore, synthesis of field emitters comprised of an array of nanostructures employing a simple and inexpensive method is a challenging task. In the present studies, we have employed a simple approach of physical masking to fabricate an array of Si-NWs on Si substrate using the HFCVD technique. Furthermore, the use of Si as a substrate is purposeful, as it facilitates integration with other solid-state electronic devices. Interestingly, the Si-NWs array emitter exhibited superior field emission behavior as compared to un-patterned emitters, delivering the maximum emission current density of ~1058.63 $\mu A/cm^2$ at an applied field of 3.9 V/μm.

6.2 Experimental Details

6.2.1 Fabrication of an Array of Si Nanowires

An array of Si-NWs was synthesized on Si substrate following the vapor–liquid-solid (VLS) principle, using Sn thin film as a catalyst. For this, an indigenously designed and commissioned cluster unit comprised of HFCVD chamber, thermal evaporation, and DC plasma systems were used. The cluster unit facilitates *in-situ* transfer of substrate between these systems. The synthesis process of Si-NWs was illustrated in Fig. 6.1a. First, a physical mask was prepared by fabricating an array of circular holes (diameter ~20 μm, distance between two holes ~400 μm) on a thin tungsten foil (1 $\times$ 1 cm, thickness ~0.1 mm) using a high-power laser. A piece of pre-cleaned Si substrate (1 $\times$ 1 cm) covered with the physical mask was mounted onto the substrate holder of the thermal evaporation unit, and Sn thin film deposition was carried out at a base pressure of ~1 $\times$ 10^{-6} mbar. The film thickness was measured using ellipsometry and found to be ~50 nm. Then the Sn thin film deposited Si substrate along with the physical mask was transferred into the DC Plasma chamber for surface cleaning. The surface cleaning was carried out using Ar plasma for 5 min duration. After the cleaning process, the Sn thin film deposited Si substrate with the physical mask was transferred into the HFCVD chamber. In order to facilitate the *in-situ* growth of Si nanowires, due to VLS, surface texturing of catalyst thin film, i.e., formation of nanometric islands, is very essential. This was achieved by thermal treatment in the presence of H_2 gas. The substrate temperature was raised to 470 °C. Prior to this, the temperature of pair of tungsten filaments was set to 1700 °C, and hydrogen (H_2) gas was introduced into the chamber using a mass flow controller (MFC) at 50 sccm. The thermal treatment was carried out for 8 min. For the HFCVD growth

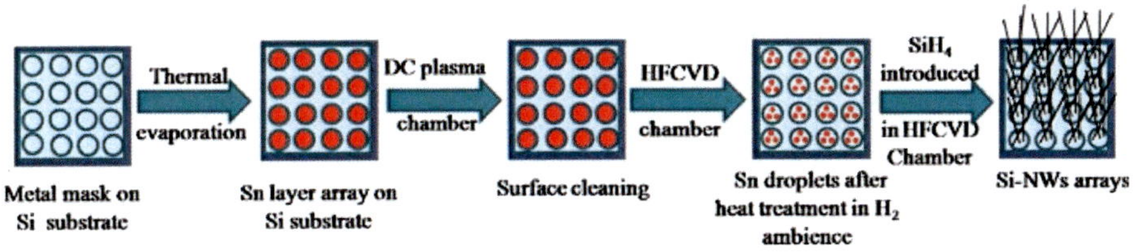

Fig. 6.1 **a** Schematic representation for preparation of array of Si-NWs

of Si nanowires, H_2 gas flow was halted, and Silane (SiH_4) was introduced into the chamber. The deposition was carried out at substrate temperatures of 470 °C, and SiH_4 flow rate was maintained at 2.5 sccm for 60 min. The entire process was done at a base pressure of ~1.2×10^{-2} mbar.

6.2.2 *Characterization Techniques*

The structural and optical properties were revealed using X-ray Diffraction (XRD, Brucker D8 Advance), Raman spectroscopy (Renishaw, UK), and UV–Visible spectrophotometer (Shimadzu UV 1800). The surface morphology and elemental analysis were revealed using FESEM (Nova NanoSEM-450 and Zeiss Ultra Plus FESEM). The field emission characteristics were investigated in a planner diode configuration at a base pressure of ~2.8×10^{-8} Torr. The as-grown Si-NWS array on the Si substrate (cathode) was pasted onto the copper holder using electroconductive silver paint. The anode, a phosphor-coated ITO glass (diameter ~50 mm) was held parallel to the cathode in close proximity (anode–cathode separation ~1000 μm). Before the FE measurements, cathode surface cleaning was attempted via in-situ ion bombardment, applying a fixed DC potential difference of ~1 kV for an hour. The FE current (I) was measured using an electrometer (Keithley 6514), whereas a high voltage DC power supply (Spellman, USA) was used to vary the applied potential difference between the anode and cathode. The FE current stability was recorded with the help of a computer-controlled data acquisition unit with the sampling time of ~10 s. In order to test the reproducibility and repeatability, the FE measurements were performed on three emitters synthesized under the same set of experimental conditions.

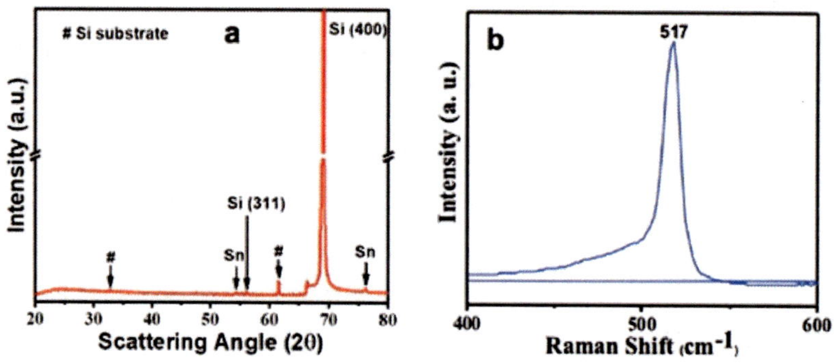

Fig. 6.2 **a** XRD pattern, and **b** Raman spectrum of un-patterned Si-NWs on Si substrate grown using HFCVD method

6.3 Results and Discussion

6.3.1 XRD and Raman Analysis

A typical X-ray diffraction pattern of un-patterned Si-NWs on Si substrate grown using the HFCVD method is depicted in Fig. 6.2a. It shows characteristic diffraction peaks indexed to (400) and (311) planes of crystalline Si-NWs. The observed diffraction peaks of the metallic Sn phase are due to the fact that Sn is used as a catalyst for the VLS growth of the Si-NWs. The crystallinity of the Si-NWs has been investigated using Raman spectroscopy. Figure 6.2b shows a typical Raman spectrum of the as-synthesized Si-NWs. The observation of a well-defined peak at ~517 cm^{-1}, which is characteristic of transverse optical Si–Si vibrational mode, clearly implies the single crystalline nature of the Si-NWs [12].

6.3.2 UV–Visible Absorption Studies

The optical absorption property of the as-synthesized Si-NWs thin film was revealed from the diffuse reflectance spectrum, shown in Fig. 6.3a. It shows maximum light absorption in the wavelength range from 490 to 780 nm. The band gap energy (Eg) was estimated from the modified Kubelka–Munk plot, i.e., plot of $F(R)(h\nu)^2$ versus energy (eV) (Fig. 6.3b), where R is the reflectance, F(R) is proportional to the extinction coefficient (α). The Eg value was found to be 1.16 eV, which is very close to the intrinsic Si bulk band gap energy (1.12 eV).

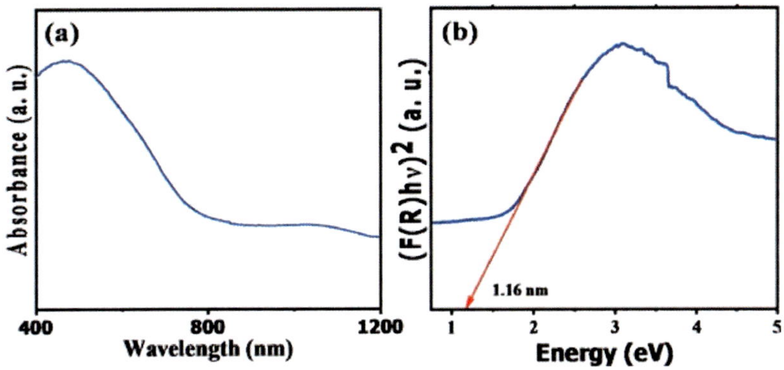

Fig. 6.3 **a** optical absorption spectrum, and **b** Kubelka–Munk plot, i.e., plot of $F(R)(h\nu)^2$ versus energy (eV) of un-patterned Si-NWs on Si substrate grown using HFCVD method

6.3.3 SEM Analysis

Figure 6.4a represents an SEM image of Sn thin film deposited on Si substrate using a physical mask. It clearly shows deposition Sn thin film comprised of an array of well-separated circular islands (of nearly the same diameter as that hole in the physical mask). The SEM image clearly justifies the success of the simple method involving the use of a physical mask to obtain a well-defined array of catalyst islands in a single-step process. The effect of thermal treatment carried out at 470 °C in H_2 environment on morphological modification of catalyst thin film was revealed by microscopic analysis. The SEM image of a single island of the catalyst thin film (Fig. 6.4b) clearly confirms the formation of nanometric size circular droplets of the catalyst upon heat treatment. The size of the catalyst droplet is very crucial in the overall growth of the Si nanostructures. Under the optimized conditions of heat treatment, the average size of catalyst droplets was found to be ~50 nm. The FESEM images of HFCVD grown Si nanostructures on Si substrate, recorded at different magnifications (Fig. 6.4c and d) clearly reveal their growth only in the regions where the catalyst islands were present. The surface morphology is characterized by randomly oriented needle-like nanowires of length ranging from ~200 to 900 nm and tip diameter between ~20 and 40 nm.

6.3.4 Field Emission Characteristics

The FE characteristics of as-synthesized Si-NWs emitter array and un-patterned Si-NWs, represented as emission current density (J) versus applied electric field (E) plot, are depicted in Fig. 6.5a. At the outset, it is clearly seen that the Si-NWs array emitter displays improved FE properties as compared to the un-patterned Si-NWs emitter. In the present study, the emission current density is defined as, $J = I/A$,

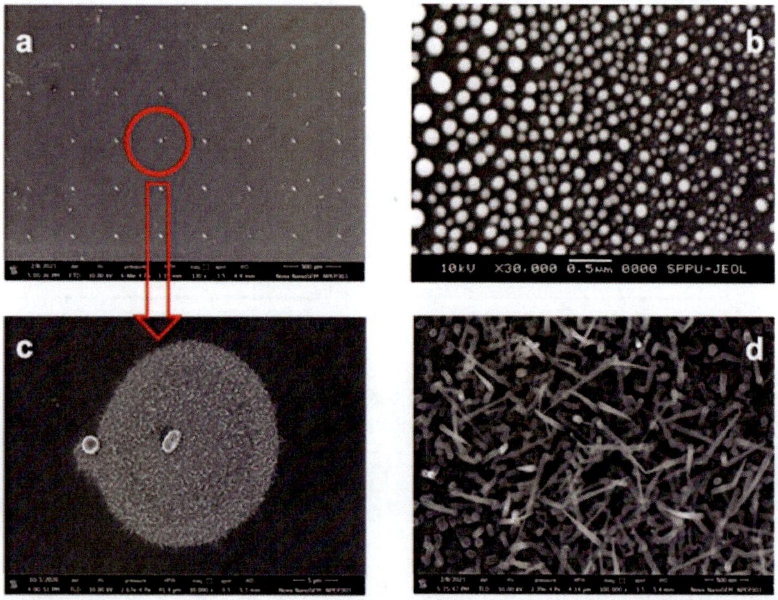

Fig. 6.4 SEM images of **a** an array of Sn thin film islands, **b** Sn droplet formation obtained upon heat treatment in H_2 environment, **c** growth of Si nanostructures only in the area where catalyst is present, and **d** growth of randomly oriented needle-like Si nanowires on Si substrate

where I is the observed emission current, and A is an area of the emitter (1 cm^2). Similarly, the applied field is defined as E = V/d, where V is the applied potential difference, and d is the separation between the anode and cathode (1 mm). The gradual increase in emission current density with the increase in the applied field in the initial region followed by the exponential increase after the turn-on field is per the Fowler–Nordheim (F-N) theory of field emission [13]. The values of the turn-on field (corresponding to an emission current density of 1 $\mu A/cm^2$), and the threshold field (corresponding to an emission current density of 10 $\mu A/cm^2$) are found to be ~1.8, and 2.2 V/μm for the Si-NWs array emitter, and ~2.1, and 2.6 V/μm for the un-patterned Si-NWs emitter, respectively. Furthermore, the emission current density of ~1058.63 $\mu A/cm^2$ has been drawn at an applied field of ~3.9 V/μm from the Si-NWs array emitter, whereas the un-patterned Si-NWs emitter delivered a maximum emission current density of ~396.6 $\mu A/cm^2$ at an applied field of ~6.8 V/μm. The values of turn-on and threshold fields exhibited by the Si-NWs array emitter are lower (superior) than the alike emitters due to well-patterned semiconductor nanostructures [14–23].

The improvement in FE characteristics exhibited by the Si-NWs array emitter, as compared to the un-patterned Si-NWs emitter, is attributed to its unique surface morphology. Both the emitters are synthesized under identical experimental conditions, except one emitter is in the form of an array of Si-NWs. A careful observation of the SEM images reveals that the areal density of Si-NWs in the array emitter is lower than that of the un-patterned emitter, and the lower the areal density, the lesser

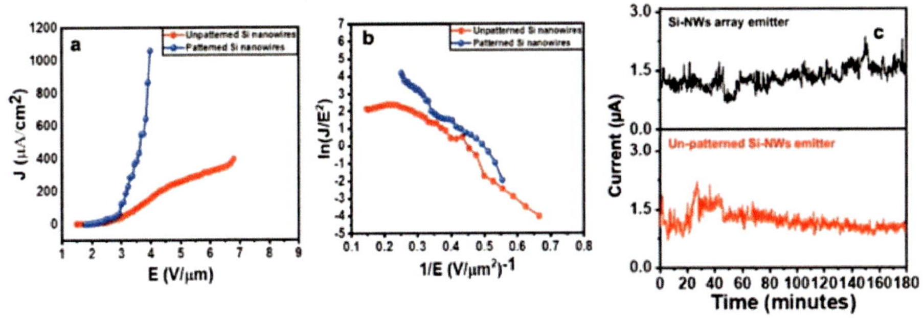

Fig. 6.5 Field emission characteristics of un-patterned and array of Si-NWs emitters **a** emission current density (J) versus applied electric field (E) behavior, **b** Fowler–Nordheim (F-N) plots, and **c** emission current stability (I-t plot) recorded at a pre-set current of 1.5 μA

the field screening, and thus betters the FE behavior. Furthermore, due to an array, it is expected that there will be a very large number of Si-NWs protruding away from the substrate, which act as potential emission sites. Thus, the improvement in FE behavior is due to unique surface morphology characterized by lower areal density (minimizing the field screening effect), with a large number of protruding Si-NWs (potential emission sites). To exemplify this, we have complied FE characteristics of alike patterned and un-patterned emitters due to semiconducting nanostructures and is presented in Table 6.1. It is very clear from Table 6.1 that the Si-NWs array emitter has superior FE behavior as compared to alike emitters.

Furthermore, the J-V characteristics were analyzed in terms of a graph of ln (J/E^2) versus (1/E), known as the F-N plot. The F-N plots corresponding to the un-patterned and array of Si-NWs emitters are shown in Fig. 6.5b. The overall nature of these F-N plots is seen to be non-linear, indicative of the semiconductive nature of the emitters. The deviation from linearity is attributed to various factors including field screening and penetration, noticeable variation in the aspect ratio of the emitters, and electronic properties (i.e., semiconducting). For semiconducting nanostructures, since electron emission occurs from both the conduction and valence bands, due to the change in effective work functions (with respect to these bands) leads to variation in the slope of the F-N plot, depicting non-linear nature.

From an application point of view, emission current stability is one of the decisive parameters. We have tested the emission current stability of both the emitters at a pre-set value of ~ 1 μA over the duration of 3 h which is depicted in Fig. 6.5c. As seen from the I-t plots, both the emitters show good stability. The emission current exhibited "spike" like fluctuations which are due to adsorption, desorption, and/or migration of residual gaseous species on the emitter surface.

Table 6.1 Compilation of the field emission characteristics of emitters due to Si nanostructures

Emitter type	Turn-on field (V/μm)	Maximum current density and respective applied field	Reference
Si Nanowires	3.35 (1 μA/cm^2)	35 μA/cm^2 at ~3.6 V/μm	15
Si Nanowires	8.5 (10 μA/cm^2)	117 μA/cm^2 at ~5.9 V/μm	20
Si Nanowires	5.5 (10 μA/cm^2)	> 1000 μA/cm^2 at ~7.0 V/μm	21
Si Nanowires	5.01 (10 μA/cm^2)	100 μA/cm^2 at ~5.9 V/μm	22
Si Nanowires	2.67 (1 μA/cm^2)	1000 μA/cm^2 at ~4.26 V/μm	23
Si Nanopillars	30.2 (200μA/cm^2)	> 1000 μA/cm^2 at ~7.0 V/μm	24
Si Nanotrees	5.20 (10 μA/cm^2)	100 μA/cm^2 at ~6.5 V/μm	25
An array of Si Nanowires	1.8 (1 μA/cm^2)	1058 μA/cm^2 at ~3.9 V/μm	Present work
Un-patterned Si Nanowires	2.1 (1 μA/cm^2)	396.6 μA/cm^2 at ~6.8 V/μm	Present work

6.4 Conclusion

A simple and single-step process dealing with physical masking has been effectively utilized to prepare an array of Silicon nanowires on Si substrate using HFCVD method with optimized process parameters. Owing to unique morphology characterized by moderate areal density offering minimal field screening effect, the Si-NWs array emitter showed improved field emission characteristics. This facile and one-step process of physical masking can be effectively extended to prepare an array of alike semiconducting one-dimensional nanomaterials.

Acknowledgements The authors (Amol Deore and Mahendra More) are grateful to the Nanomission, Department of Science and Technology (DST), Govt. of India for the financial assistance to commission the HFCVD cluster unit and fellowship [grant code-SR/NM/NS-52/2016). Krishna Jagtap would like to acknowledge the financial support as a Research Associate from CEPIFRA, DST, Govt. of India, (Project 62T8-1). The field emission work was carried out as a part of the CEPIFRA project.

References

1. Zeng, B., Xiong, G., Chen, S., Wang, W., Wang, D.Z., Ren, Z.F.: Field emission of silicon nanowires grown on carbon cloth. Appl. Phys. Lett. **90**(3), 033112 (2007)
2. Feng, S.Q., Yu, D.P., Zhang, H.Z., Bai, Z.G., Ding, Y.: The growth mechanism of silicon nanowires and their quantum confinement effect. J. Cryst. Growth **209**(2–3), 513–517 (2000)
3. Chan, C.K., Peng, H., Liu, G., McIlwrath, K., Zhang, X.F., Huggins, R.A., Cui, Y.: High-performance lithium battery anodes using silicon nanowires. Nat. Nanotechnol. **3**(1), 31–35 (2008)
4. Demami, F., Ni, L., Rogel, R., Salaun, A.C., Pichon, L.: Silicon nanowires based resistors as gas sensors. Sens. Actuators, B Chem. **170**, 158–162 (2012)
5. Tian, B., Zheng, X., Kempa, T.J., Fang, Y., Yu, N., Yu, G., Lieber, C.M.: Coaxial silicon nanowires as solar cells and nanoelectronic power sources. Nature **449**(7164), 885–889 (2007)
6. Riccitelli, R., Di Carlo, A., Fiori, A., Orlanducci, S., Terranova, M.L., Santoni, A., Villacorta, F.J.: Field emission from silicon nanowires: conditioning and stability. J. Appl. Phys. **102**(5), 054906 (2007)
7. Zeng, B., Xiong, G., Chen, S., Jo, S. H., Wang, W. Z., Wang, D. Z., Ren, Z. F.: Field emission of silicon nanowires. Appl. Phys. Lett. **88**(21), 213108 (2006)
8. Kichambare, P.D., Tarntair, F.G., Chen, L.C., Chen, K.H., Cheng, H.C.: Enhancement in field emission of silicon microtips by bias-assisted carburization. J. Vac. Sci. & Technol. B: Microelectron. Nanometer Struct. Process., Meas., Phenom. **18**(6), 2722–2729 (2000)
9. Mu, C., Yu, Y., Liao, W., Zhao, X., Xu, D., Chen, X., Yu, D.: Controlling growth and field emission properties of silicon nanotube arrays by multistep template replication and chemical vapor deposition. Appl. Phys. Lett. **87**(11), 113104 (2005)
10. Basu, A., Swanwick, M.E., Fomani, A.A., Velásquez-García, L.F.: A portable x-ray source with a nanostructured Pt-coated silicon field emission cathode for absorption imaging of low-Z materials. J. Phys. D Appl. Phys. **48**(22), 225501 (2015)
11. Djuzhev, N.A., Demin, G.D., Filippov, N.A., Evsikov, I.D., Glagolev, P.Y., Makhiboroda, M.A., Bespalov, V.A.: Development of technological principles for creating a system of microfocus X-ray tubes based on silicon field emission nanocathodes. Tech. Phys. **64**(12), 1742–1748 (2019)
12. Tarun, A., Hayazawa, N., Ishitobi, H., Kawata, S., Reiche, M., Moutanabbir, O.: Mapping the "forbidden" transverse-optical phonon in single strained silicon (100) nanowire. Nano Lett. **11**(11), 4780–4788 (2011)
13. Fowler R.H., Nordheim L.: Electron emission in intense electric fields. In: Proceedings of the royal society of London. Series A, containing papers of a mathematical and physical character 119, pp. 173–181. The Royal Society Publishing (1928)
14. Huang, Y.H., Lin, H.C., Cheng, S.L.: Fabrication of vertically well-aligned $NiSi_2$ nanoneedle arrays with enhanced field emission properties. J. Phys. Chem. Solids **150**, 109892 (2021)
15. Lv, S., Li, Z., Liao, J., Wang, G., Li, M., Miao, W.: Optimizing field emission properties of the hybrid structures of graphene stretched on patterned and size-controllable SiNWs. Sci. Rep. **5**(1), 1–8 (2015)
16. Ahsanulhaq, Q., Kim, J.H., Hahn, Y.B.: Controlled selective growth of ZnO nanorod arrays and their field emission properties. Nanotechnology **18**(48), 485307 (2007)
17. Sohn, J.I., Lee, S., Song, Y.H., Choi, S.Y., Cho, K.I., Nam, K.S.: Patterned selective growth of carbon nanotubes and large field emission from vertically well-aligned carbon nanotube field emitter arrays. Appl. Phys. Lett. **78**(7), 901–903 (2001)
18. Yang, W., Wang, W., Xie, C.: Large scale growth of patterned SiC nanowire arrays and their field emission performance. J. Am. Ceram. Soc. **102**(7), 3854–3859 (2019)
19. Chubenko, O., Baturin, S. S., Kovi, K. K., Sumant, A. V., Baryshev, S.V.: Locally resolved electron emission area and unified view of field emission from ultra-nanocrystalline diamond films. ACS Appl. Mater. & Interfaces **9**(38), 33229–33237 (2017)

20. Devarapalli, R.R., Kashid, R.V., Deshmukh, A.B., Sharma, P., Das, M.R., More, M.A., Shelke, M.V.: High efficiency electron field emission from protruded graphene oxide nanosheets supported on sharp silicon nanowires. J. Mater. Chem. C **1**(33), 5040–5046 (2013)
21. Cheng, S.L., Lin, H.C., Huang, Y.H., Yang, S.C.: Fabrication of periodic arrays of needle-like Si nanowires on (001) Si and their enhanced field emission characteristics. RSC Adv. **7**(39), 23935–23941 (2017)
22. Zhao, F., Cheng, G.A., Zheng, R.T., Zhao, D.D., Wu, S.L., Deng, J.H.: Field emission enhancement of Au-Si nano-particle-decorated silicon nanowires. Nanoscale Res. Lett. **6**(1), 1–5 (2011)
23. Hung, Y.J., Lee, S.L., Beng, L.C., Chang, H.C., Huang, Y.J., Lee, K.Y., Huang, Y.S.: Relaxing the electrostatic screening effect by patterning vertically-aligned silicon nanowire arrays into bundles for field emission application. Thin Solid Films **556**, 146–154 (2014)
24. Chang, Y.M., Ravipati, S., Kao, P.H., Shieh, J., Ko, F.H., Juang, J.Y.: Broadband antireflection and field emission properties of TiN-coated Si-nanopillars. Nanoscale **6**(16), 9846–9851 (2014)
25. Lv, S., Li, Z., Chen, C., Liao, J., Wang, G., Li, M., Miao, W.: Enhanced field emission performance of hierarchical ZnO/Si nanotrees with spatially branched heteroassemblies. ACS Appl. Mater. Interfaces. **7**(24), 13564–13568 (2015)

Chapter 7
Photoluminescence Property of Nano Silica Mixed YAG:Ce Phosphors

Tatsuki Sogabe, Takaaki Sakai, Satoshi Hiroi, Koji Ohara, Satoshi Sugano, Shao-Ju Shih, Toshihiro Moriga, and Masatsugu Oishi

Abstract Typical white light-emitting diode (LED) lighting is made by combining blue LED for excitation and yellow phosphor, which absorb blue light and emit yellow light. One way to further improve the performance of white LED lighting is to improve the quantum efficiency of the yellow phosphor. A common method to improve the quantum efficiency of phosphors is to coat the phosphor surface with an oxide material. Therefore, increasing the surface area of the phosphors, followed by mixing with nano-silica, would be effective in further improving the quantum efficiency of the phosphors. In this study, yttrium aluminum garnet (YAG) powder, which is widely used as a yellow phosphor for white LED lighting, was pulverized by ball milling and composite YAG powders mixed with nano-silica prepared in a tetraethyl orthosilicate (TEOS) solution at room temperature was evaluated. We obtained smaller particles by ball milling. In addition, the composite YAG powders were successfully fabricated without changing the emission spectrum of YAG, but quantum efficiency was unchanged or decreased due to the non-uniform silica powders on YAG particles.

T. Sogabe · S. Sugano · T. Moriga · M. Oishi (✉)
Graduate School of Technology, Industrial and Social Sciences, Tokushima University, 2-1 Minami Josanjima-cho, Tokushima 770-8506, Japan
e-mail: ooishi.masatsugu@tokushima-u.ac.jp

T. Sakai
Global Zero Emission Research Center, National Institute of Advanced Industrial Science and Technology (AIST), 16-1 Onogawa, Tsukuba, Ibaraki 305-8569, Japan

S. Hiroi · K. Ohara
Diffraction and Scattering Division, Japan Synchrotron Radiation Research Institute (JASRI), 1–1–1 Kouto, Sayo-cho, Sayo-gun, Hyogo 351-0198, Japan

S.-J. Shih
Department of Materials Science and Engineering, National Taiwan University of Science and Technology, Taipei 10607, Taiwan

R. Murakami et al. (eds.), *The 3rd International Conference on Nanomaterials and Advanced Composites*, Springer Proceedings in Physics 298,
https://doi.org/10.1007/978-981-99-7153-4_7

7.1 Introduction

Solid lighting using a light-emitting diode (LED) is more energy-efficient than incandescent and fluorescent lamps, hence it is widely used as an environmentally friendly lighting source [1]. Typical white LED lighting is composed by combining a blue LED for excitation and a yellow phosphor that absorbs the blue light and emits yellow light. A pseudo-white color is obtained by mixing blue and yellow emissions. Therefore, improving the quantum efficiency of the yellow phosphor is considered to be effective in further reducing the energy consumption of white LED lighting. A common method to improve the quantum efficiency of phosphors is to coat the phosphor surface with an oxide material [2, 3]. We reported the composite red phosphor $CaAlSiN_3$:Eu mixed with nano-sized silica prepared in a tetraethyl orthosilicate (TEOS) solution in an alkaline hydrolysis bath, which quantum efficiency (QE) improved by about 15% [4]. This is attributed to the improvements in the light extraction efficiency of phosphor particles. Therefore, a larger surface area of the phosphor followed by mixing nano-sized silica would be valid to further improve the QE of phosphors. Yttrium aluminum garnet (YAG) powders, which is widely used as a yellow phosphor for white LED illumination, were pulverized using ball milling and then mixed with nano-silica prepared in TEOS solution at room temperature. The surface morphology of the particles was evaluated by field emission scanning electron microscopy (FE-SEM); the crystal structure was analyzed by X-ray diffraction (XRD) and pair distribution function (PDF), and the photoluminescence properties were measured. We report the photoluminescence properties of pulverized YAG powder and the composite YAG powder.

7.2 Experimental

Commercialized YAG phosphor $Y_3Al_5O_{12}$:Ce powder (Mitsubishi Chemical Corporation) was used. YAG powders were pulverized using the ball milling machine (Fritsch Pulverizette 7, Fritsch). The powder was ball milled using φ15 mm, φ2 mm, and φ0.05 mm zirconia beads in ethanol. The YAG powder was first pulverized at 300 rpm for 1 h using φ15 mm balls, then, at 500 rpm for 1 h using φ2 mm balls, and finally at 800 rpm for 7 h using φ0.05 mm balls [5, 6]. For the composite YAG powders, ball-milled YAG powder was dispersed in isopropanol (C_3H_8O, 99.7%, Wako), and the TEOS ($Si(OC_2H_5)_4$, 95%, Wako) solution. This solution was stirred ultrasonically for 1 min. Aqueous ammonia (NH_3, 28–30%, Wako) solution was added to initiate the polymerization reaction [7, 8]. The solution was stirred at a rate of 500 rpm using a magnetic stirrer and the temperature of the solutions was maintained at room temperature throughout the synthesis. Two samples were prepared with different amounts of aqueous ammonia and different stirring times. For composite YAG (2ml_2h), 2 ml of 18% ammonia was mixed and stirred for 2 h. For composite YAG (0.5ml_4h), 0.5 ml of 18% ammonia was mixed and stirred for 4 h. The crystal structure of samples was evaluated by XRD (Miniflex, Rigaku). The powder morphology was observed by scanning electron microscopy (SEM) (JSM-6510, JEOL), and field emission scanning electron microscopy (FE-SEM) (S-4700, Hitachi High-tech). The QE and absorbance (ABS) were measured using an absolute quantum yield spectrometer (Quantaurus-QY C11347-01, Hamamatsu Photonics). The high-energy X-ray total scattering experiments were performed at room temperature at the SPring-8 beam-line BL04B2 using a two-axis diffractometer [9]. The incident X-ray energy obtained from a Si(220) crystal monochromator was 61.4 keV. The diffraction patterns of the samples and an empty tube were measured in the transmission geometry with an angle of 0.3–49°, corresponding to a Q-range of 0.2–25 Å^{-1}. The pair distribution function (PDF) profiles were obtained using the in-house BL04B2 software. The X-ray structure factor, $S(Q)$, was derived from the diffraction intensity. Then the reduced PDF, $G(r)$, was obtained by a Fourier transformation of $S(Q)$, which is given as

$$G(r) = \frac{2}{\pi}\int_{Q_{\min}}^{Q_{\max}} Q\{\mathrm{S}(Q) - 1\}\sin(Q\mathrm{r})\mathrm{d}Q, \tag{7.1}$$

where $Q_{\max}$ and $Q_{\min}$ are the maximum and minimum observed Q values, respectively.

7.3 Results and Discussion

Figure 7.1 shows the XRD profiles of the pure and ball-milled YAG powders. The XRD profiles are the same after the ball millings, indicating that the crystal structure of YAG particles did not change by the ball milling processes. For the powders ball milled with φ0.05 mm balls, the peaks were broadened, and the background increased compared to the other samples. These results suggest the distortion of crystal structure or formation of nano-sized particles.

Figure 7.2 shows the SEM images of pure YAG powder before and after the ball milling and composite YAG powders. Figure 7.2a~c was observed by SEM and Fig. 7.2d~g by FE-SEM, respectively. The pure YAG particles are about 10 μm in size, the particle surface is smooth, and its shape is spherical (Fig. 7.2a). After ball milling with φ15 mm balls, the YAG particles have an average size of 5 μm and the surface is smooth and sharp (Fig. 7.2b). After ball milling with φ2 mm balls, the YAG particle size did not change compared to that of φ15 mm balls, but the particle size appeared to be more uniform (Fig. 7.2c). After the ball milling with φ0.05 mm balls, the YAG particles are also mostly unchanged (about 5 μm in size), but the particle surface is no longer smooth (Fig. 7.2d). This is due to the agglomeration of YAG particles that were refined by ball milling and behaved as one large particle. After the particles were dispersed by supersonic waves, we observed many YAG particles of about 1 μm in diameter (Fig. 7.2e).

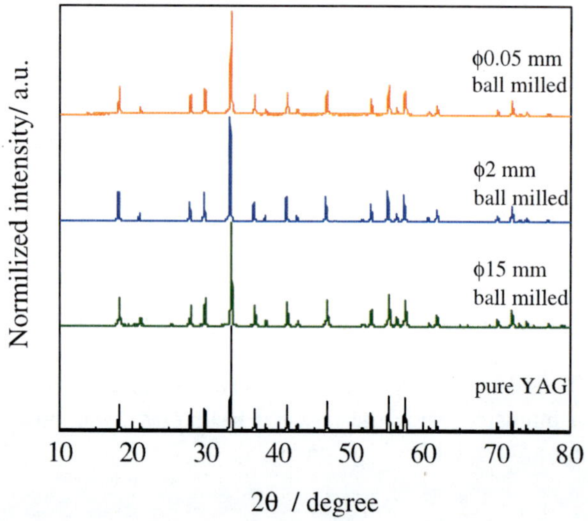

Fig. 7.1 XRD profiles of YAG powders

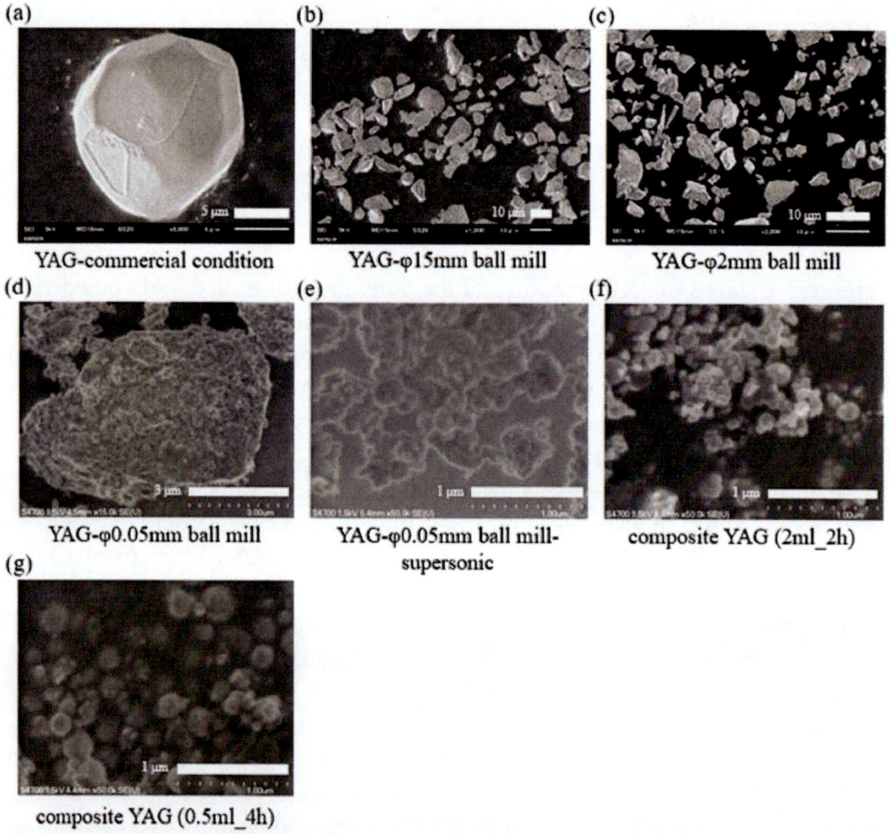

Fig. 7.2 SEM images of YAG powders

The composite YAG powders were prepared in two different synthesis conditions in TEOS solution. From SEM observations, we observed spherical particles of about 100 nm in diameter (Fig. 7.2f and g). These microscopic spherical particles were not observed for ball-milled YAG samples; hence, these round particles are SiO_2 particles. Regarding the surface of the microscopic spherical particles, they are smooth in composite YAG (2ml_2h), but not in composite YAG (0.5ml_4h). This indicates that the composite YAG particles of composite YAG (0.5ml_4h) are covered by SiO_2 particles.

The distributions of particle sizes of YAG particles after ball milling are shown in Fig. 7.3. 100 particles were randomly selected, and the particle sizes were measured. Figure 7.3a shows that the ball milling changed the particle size distribution of the YAG particles to the micro-region. Figure 7.3b compares the particle size before and after the ultrasonic dispersion of φ0.05 mm ball-milled particles. Single micro-sized particle increases after the ultrasonic dispersion, which indicates the agglomeration of YAG particles after ball milling with φ0.05 mm balls.

The picture of pure YAG and obtained composite YAG powders are shown in Fig. 7.4. White-colored powder was obtained for composite YAG (2ml_2h), which is different from that of the pure YAG. Hence, the composite YAG (2ml_2h) sample is mostly composed of SiO_2 powders. On the other hand, composite YAG (0.5ml_4h) was yellowish but lighter in color than the pure YAG. Hence, it suggests a mixture of SiO_2 and YAG particles.

$S(Q)$ profiles of pure YAG and composite YAG powders are shown in Fig. 7.5a. Composite YAG (0.5ml_4h) showed strong diffraction peaks corresponding to those of YAG powders. While for composite YAG (2ml_2h), it showed broad peaks representing amorphous SiO_2. Strong peaks representing YAG powder are also recognizable suggesting small contents of YAG particles in composite YAG (2ml_2h). The $G(r)$ profiles at short-range distances are shown in Fig. 7.5b. For composite YAG (0.5ml_4h), in addition to the correlations of amorphous SiO_2 structures, small correlation peaks corresponding to those of YAG are also recognized. The ratio of YAG particles to SiO_2 particles was calculated from $G(r)$ profiles. For composite YAG (2ml_2h), YAG:SiO_2 ratio is 8:92 mol%, which suggests a large portion of SiO_2 particles. While for composite YAG (0.5ml_4h), YAG:SiO_2 ratio is 48:52 mol%. It suggests that the reaction condition has a strong influence on the formation of SiO_2 particles.

Figure 7.6a shows the emission spectra measured at the excitation wavelength at 460 nm. The emission spectra were unchanged after the pulverization and for composite YAG samples. Figure 7.6b and c shows the ABS and QE. ABS increased for ball-milled YAG powders. This is due to an increase in the specific surface area of the milled particles, or higher packing powder density of the ball-milled YAG particles [10]. For the composite YAG powders, ABS decreased due to the reflection by the mixed SiO_2 particles. QE decreased for pulverized YAG powders and composite YAG powders. It may be attributed to the increase in luminescence killers originating from the surface of the YAG particles [11]. Due to the increased specific surface area of pulverized particles, the effects of luminescence killers could be enhanced. For the composite YAG powders, we expected the increase of QE

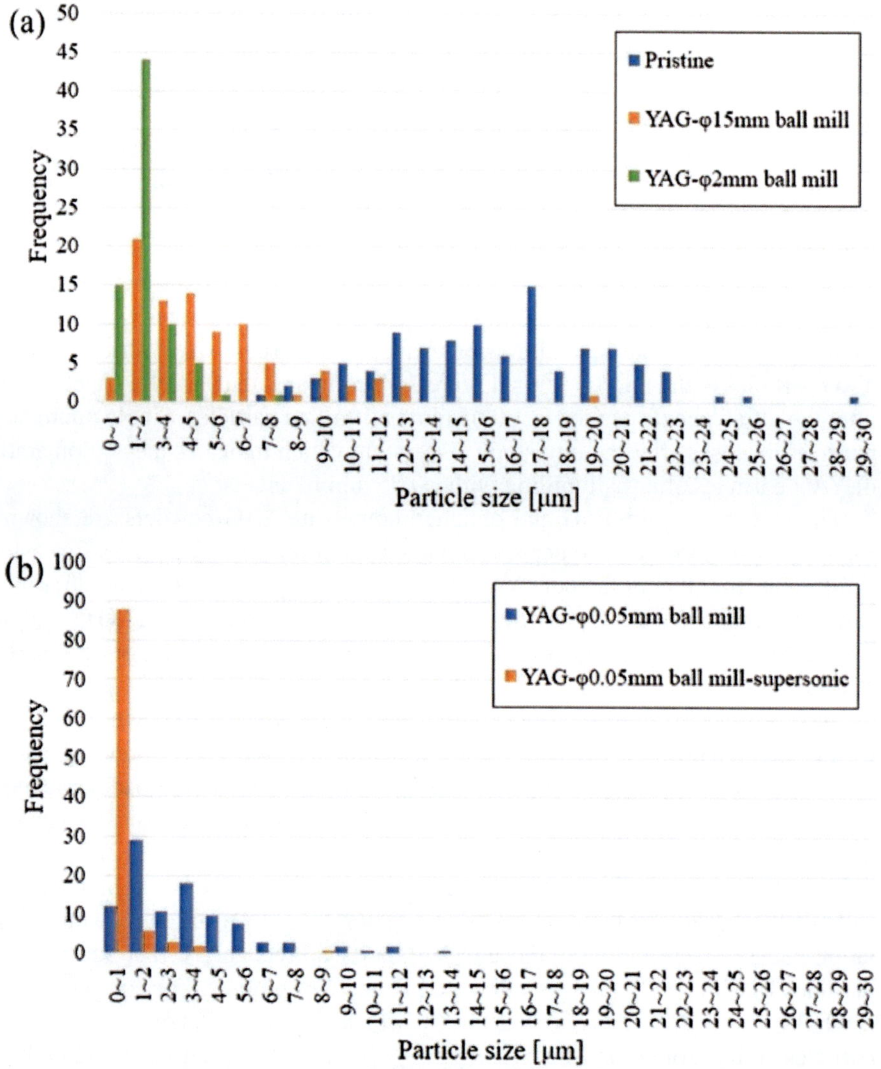

Fig. 7.3 **a** Particle size distributions of before and after ball millings and **b** comparison before and after ultrasonic treatment of φ0.05 mm ball-milled particles

by the increased light emission efficiency but this was not observed for the samples evaluated in this study. This may be due to the inhomogeneous SiO_2 particles formed on YAG particles. To improve the QE of composite YAG powders, coating with a thin SiO_2 layer would be effective [12]. We are currently working to find the appropriate TEOS bath conditions to obtain these composite YAG powders.

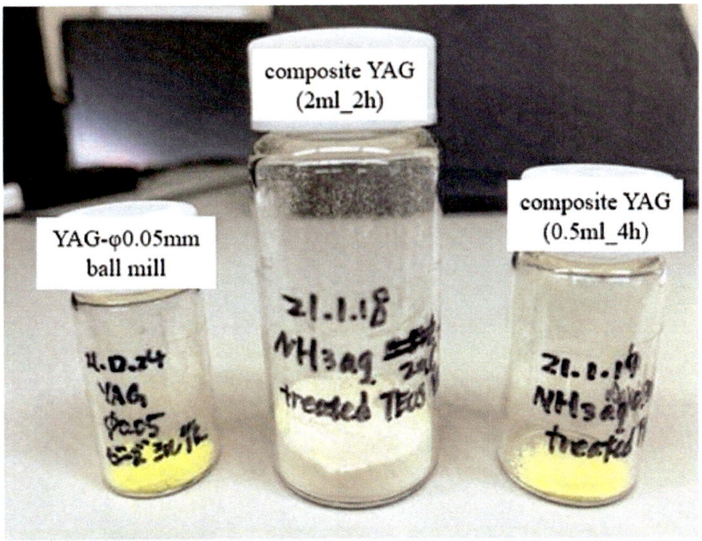

Fig. 7.4 Pure YAG powders and composite YAG powders obtained in TEOS solution

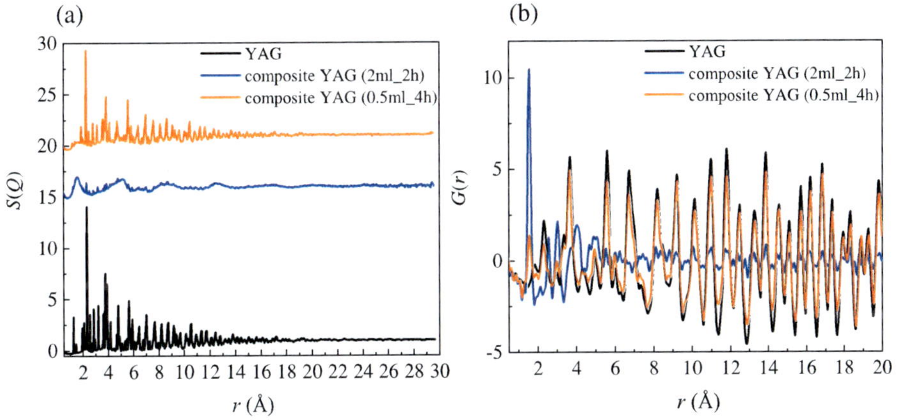

Fig. 7.5 **a** $S(Q)$ and **b** $G(r)$ profiles of pure YAG and two composite YAG powders

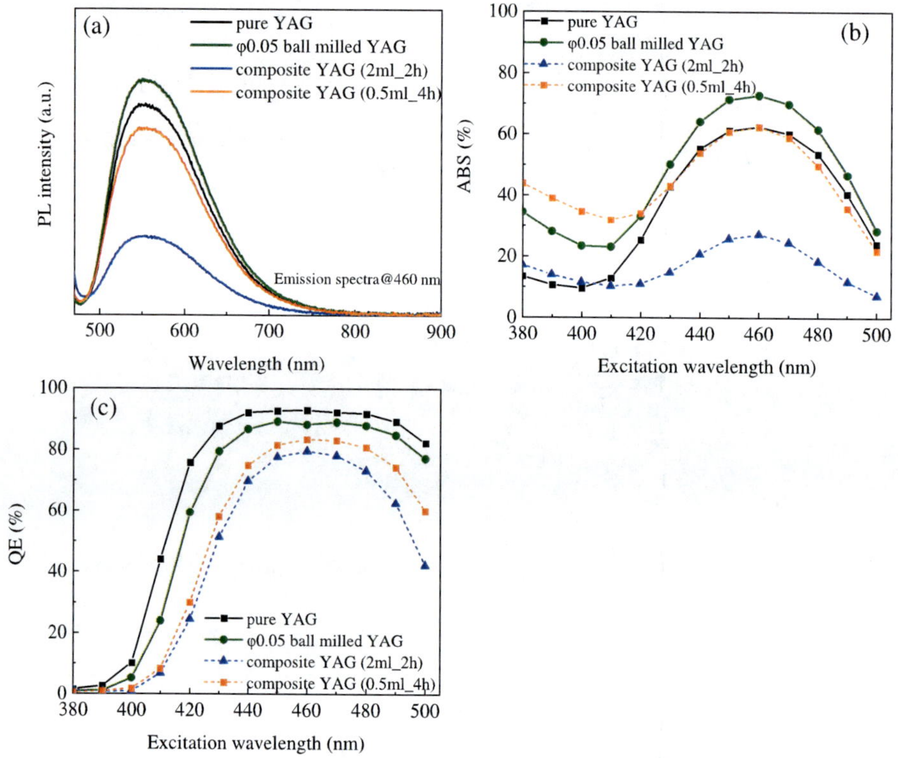

Fig. 7.6 **a** Emission spectra, **b** ABS, and **c** QE of the YAG samples

7.4 Conclusion

We succeeded in pulverizing YAG particles using ball milling. The composite YAG powders were synthesized in TEOS solution at room temperature. Appling smaller amounts of ammonia water as the initiator and catalysis, and a reaction time of 4 h, we obtained a composite powder of YAG and SiO_2 particles. The emission characteristics did not change between the pure and composite YAG samples; however, the QE of pulverized and composite YAG powders decreased compared to that of pure YAG. This is due to the surface defects formed on the YAG surface by the ball milling process, and ununiform SiO_2 particles formed on the YAG particles.

Acknowledgements We thank the Technology Center for Regional R & D, Tokushima University for the use of the facilities. This work was performed under the Collaboration Program by and between Tokushima University (TU) and the National Taiwan University of Science and Technology (NTUST). The X-ray total scattering measurements were carried out at SPring-8 under proposals 2019A1141, 2019B1148, and 2022A2083.

References

1. Taguchi, T., Member.: Present status of energy saving technologies and future prospect in white LED lighting. IEEJ Trans. Electr. Electron. Eng. **3**, 21–26 (2008)
2. Sohn, S.H., Lee, J.H., Lee, S.M.: Effects of the surface coating of $BaMgAl_{10}O_{17}:Eu^{2+}$ phosphor with SiO_2 nano-particles. J. Lumin. **129**, 478–481 (2009)
3. Cervantes-Vásquez, D., Contreras, O.E., Hirata, G.A.: Quantum efficiency of silica-coated rare-earth doped yttrium silicate. J. Lumin. **143**, 226–232 (2013)
4. Oishi, M., Shiomi, S., Ohara, K., Fujishiro, F., Kai, Y., Shin, S.-J., Moriga, T., Chichibu, S.F., Takatori, A., Kojima, K.: Enhanced quantum efficiency of a self-organized silica mixed phosphor $CaAlSiN_3$. J. Solid State Chem. **309**, 122968 (2022)
5. Sakai, T., Hyodo, J., Ishihara, T., Matsumoto, H.: Single Nanosize Pulverization of solid oxide by means of a wet planetary-beads-milling. J. Ceram. Soc. Jpn. **120**, 39–42 (2012)
6. Sakai, T., Matsumoto, H., Sato, Y., Hyodo, J., Ito, N., Hashimoto, S., Ishihara, T.: High sinterabillity of planetary-bead-milled barium zirconate. Electrochem **77**, 876–878 (2009)
7. Shiomi, S., Kawamori, M., Yagi, S., Matsubara, E.: One-pot synthesis of silica-coated copper nanoparticles with high chemical and thermal stability. J. Colloid Interface Sci. **460**, 47–54 (2017)
8. Shiomi, S., Matsubara, E., Taguchi, H., Hashida, S., Yokohama, T.: Synthesis of silicacoated copper nanoparticles and its application to red color glaze. Adv. Mater. Res. **970**, 288–292 (2014)
9. Ohara, K., Onodera, Y., Murakami, M., Kohara, S.: Structure of disordered materials under ambient to extreme conditions revealed by synchrotron x-ray diffraction techniques at SPring-8 recent instrumentation and synergic collaboration with modelling and topological analysis. J. Phys.: Condens. Matter **33**, 383001 (2021)
10. Hyun, K.Y., Jong, W.C., Byung, K.M., Jung, H.J., Ki-wan, J., Ho, S.L., Soung, S.Y.: Photoluminescence investigations of YAG: Eu nanocomposite powder by high-energy ball milling. Curr. Appl. Phys. **9**, e86–e88 (2009)
11. Kasuya, R., Isobe, T., Kuma, H.: Glycothermal synthesis and photoluminescence of YAG: Ce^{3+} nanophosphors. J. Alloy. Compd. **408–412**, 820–823 (2016)
12. Yahui, S., Xinqing, S., Haitao, Z., Renli, F., Qinjiang, H., Min, Z., Haidong, R.: Simulation of optical behavior of YAG:Ce^{3+}@SiO_2 phosphor used for chip scale packages WLED. J. Lumin. **244**, 118699 (2022)

Chapter 8
In Situ Observation of Crystal Growth Processes

Yoshihisa Suzuki, Ai Ninomiya, and Shinichiro Yanagiya

Abstract In situ observation is a powerful method for a deeper understanding of the fundamentals of crystal growth. In this review, we introduce recent advancements in protein crystallization and colloidal crystallization. From the advancement rates of molecular steps on the growth interface of highly purified protein crystals, we have calculated the vibration frequency of protein molecules adsorbed on the crystal surface for the first time. This was conducted in situ as a part of the international space station experiments under microgravity conditions. We also observed in situ the growth interface of colloidal crystals formed with depletion attractive interactions between polystyrene particles. We successfully measured the "peeling off" processes induced by the adsorption of a particle from bulk solution on a particle on a step edge of the growth interface for the first time.

8.1 Introduction

Crystal growth processes are always very important for the improvement of the quality of the obtained crystals for several devices [1]. Precise control of crystal growth processes is key to obtaining high-quality crystals. High-quality crystals sometimes make groundbreaking technical improvements. For instance, high-quality single crystals of neuraminidase are needed for the synthesis of effective antiviral inhibitors and dramatic improvements in flu medications [2].

On the other hand, the growth of high-quality crystals is always difficult in many fields. Although the drug design based on the precise structure of proteins often requires a diffraction resolution of crystals higher than 1.5 Å for the reduction of

Y. Suzuki (✉) · A. Ninomiya
Graduate School of Technology, Industrial and Social Sciences, Tokushima University, 2-1 Minamijosanjima, Tokushima 770-8506, Japan
e-mail: yoshis@tokushima-u.ac.jp

S. Yanagiya
Institute of Post-LED Photonics, Tokushima University, 2-1 Minamijosanjima, Tokushima 770-8506, Japan

R. Murakami et al. (eds.), *The 3rd International Conference on Nanomaterials and Advanced Composites*, Springer Proceedings in Physics 298,
https://doi.org/10.1007/978-981-99-7153-4_8

the difficulty of determining the target proteins' structure [3], the Protein Data Bank showed that the X-ray resolution of only about 10% of structures was higher than 1.5 Å.

In situ observation of crystal growth processes would clarify precise kinetics and give us appropriate routes to approach high-quality crystals. Lateral growth rates of hen egg-white lysozyme crystals measured in situ by Michelson interferometry in the space station took almost the same values as those on the Earth, whereas the crystal growth rates under microgravity had been conventionally expected to be suppressed compared to those on the Earth owing to the lack of gravitational convection around crystals growing in space [4].

In this review, we would like to introduce recent advancements in fundamental understandings of protein crystallization and colloidal crystallization, which have been studied via in situ observation as follows. (1) From the advancement rates of molecular steps on the growth interface of glucose isomerase (GI) crystals grown from 99.9% pure GI solutions, we calculated the vibration frequency of GI molecules adsorbed on the crystal surface for the first time as a part of the Advanced Nano Step (AdNano) mission conducted in the KIBO on the international space station [5]. (2) We also observed in situ the growth interface of colloidal crystals formed with depletion attractive interactions between polystyrene particles. We successfully measured as we say peeling-off process induced by another adsorption of a particle from bulk solution on a particle on a step edge of the growth interface [6]. Such phenomena cannot be anticipated without in situ observation.

8.2 Vibration Frequency of GI Molecules Adsorbed on the Crystal Growth Interface

8.2.1 In Situ Observation

Clear spiral growth hillocks on the {110} face of GI crystals grown on the chemically fixed seed GI crystals with straight steps and sharp edges were observed in situ both on the ground and under microgravity conditions (Fig. 8.1) [5]. Here, a spiral hillock is defined as the spiral-shaped sequential step structure formed around a screw dislocation in a crystal. The spiral hillocks around screw dislocations were generated owing to the lattice mismatches between the chemically fixed GI seed crystals and raw GI crystals regrown over the seeds, the roughening of the interface of the seed, and the induced lattice distortions [7]. Using the micrographs, we measured step velocities V_{step} and lateral growth rates V_{lateral} toward the same crystallographic direction.

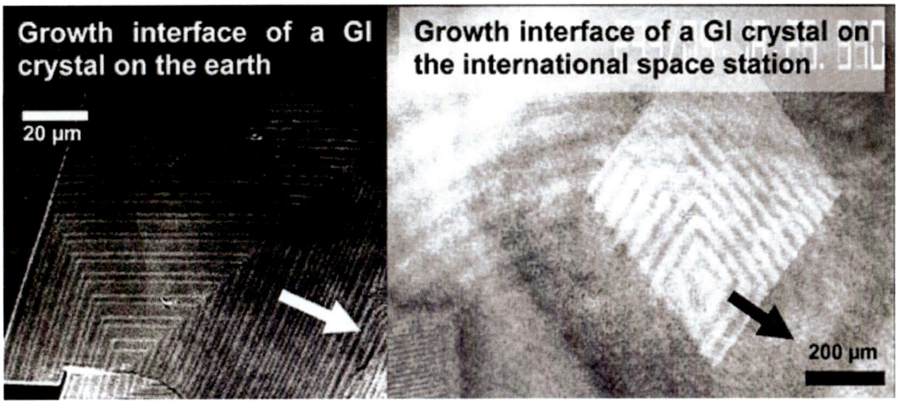

Fig. 8.1 Micrographs of the {110} faces of GI crystals on the ground (left) and under microgravity (right). A spiral growth hillock of elementary steps is clearly observed using a laser confocal microscope combined with a differential interference contrast microscope (LCM-DIM) in the left picture. The white arrow indicates the direction in which V_{step} was measured. The scale bar represents 20 μm. In the right micrograph, each parallelogram-shaped interference fringe on the {110} face indicates the same altitudes of a spiral growth hillock. The black arrow indicates the direction of the measured V_{lateral}. The scale bar represents 200 μm. Reprinted with permission from the reference [5]. Copyright 2022 American Chemical Society

8.2.2 *Step Velocities* $\mathbf{V}_{step}$ *and Lateral Growth Rates* $\mathbf{V}_{lateral}$

We measured V_{step} and V_{lateral} with temperature and plotted them with $C-C_{\text{e}}$ (C: GI concentration and C_{e}: GI solubility) (Fig. 8.2). C in ground-based experiments (for V_{step}) was measured with an ultraviolet and visible light spectrometer [8]. We measured the equilibrium temperature T_{e} at C via in situ observation of the back-and-forth movements of elementary steps on the earth. C_{e} at each temperature was calculated for both ground and space experiments using a linear function fitted to Van't Hoff plots of ln C versus $1/T_{\text{e}}$. Over the AdNano mission, T_{e} was measured in situ by observing interference fringe movements. With the use of the above-mentioned linear function, we calculated C at T_{e}. Figure 8.2 shows V_{step} and V_{lateral}. Close and open circles indicate V_{step} and V_{lateral}, respectively. V_{step} and V_{lateral} had similar values up to $C-C_{\text{e}} = 10$ mg mL^{-1}, whereas an apparent drastic increase in V_{lateral} at larger $C-C_{\text{e}}$ values owing to two-dimensional nucleation between adjacent spiral steps was found. The first result is against the existing hypothesis that growth rates of crystals under microgravity should be smaller than those on the earth owing to no convection flows around the crystals and the decrease in solute transportation onto the growth interfaces. In situ observation of the morphology of growth interfaces and the concentration gradients around the crystal achieved to solve the long-time misunderstandings. In the case of diffusion-limited growth assumed in the conventional hypothesis, the concentration just at the growth interface of the crystal under microgravity should be much smaller than that at 1 G because of the convection agitation of the solution around the growing crystal. Actually, the decrease in

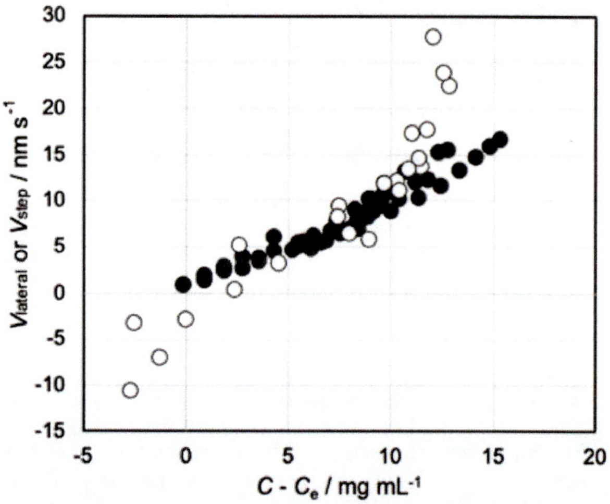

Fig. 8.2 V_{step} on the earth (●) and $V_{lateral}$ under microgravity (○). Reprinted with permission from the reference [5]. Copyright 2022 American Chemical Society

the concentration just at the growth interface of a GI crystal under microgravity conditions was very small. The growth of the GI crystal under microgravity was not diffusion-limited but interfacial-reaction-limited.

8.2.3 *Vibration Frequency of a GI Tetramer*

V_{step} or $V_{lateral}$ is approximately expressed as [5]

$$V_{step} \text{or} V_{lateral} = \beta_{step} \Omega (C - C_e)$$

where, β_{step}: the step kinetic coefficient, Ω: the volume of a GI tetramer in a GI crystal. β_{step} is further expressed as

$$\beta_{step} = \nu a \left(\frac{s}{\lambda}\right) \exp(-\frac{\varepsilon_{kink}}{kT})$$

where, ν: the vibration frequency of the GI tetramer at a kink site on the step, s: the length of the tetramer parallel to a step, a: the length of the tetramer normal to the step, λ: kink spacing on the step, and ε_{kink}: the activation energy of kink incorporation of a GI tetramer. Here "kink" is defined as the site on a step where particles are incorporated into a crystal. To calculate the vibration frequency, first, constants s and a are substituted in the above equation. Second, ε_{kink} was assumed to be zero. Nonzero ε_{kink} results in the changes in β_{step} with temperature, whereas β_{step} was almost constant even though plots of the ground experiments (●) was measured by changing temperatures. Finally, λ was calculated to be from 3.5 s to 18 s: the maximum λ was calculated to be 18 s from the interference fringe photos, and the minimum one should be larger than 3.5 s [10]. $\lambda = 18s$ was estimated from Fig. 8.1

as follows. The bottom-right side of a parallelogram interference fringe in Fig. 8.1 looks almost straight. Its direction slightly deviates from the direction parallel to the bottom-right edge of the {110} face of the GI crystal. If the deviation was assumed to be fully owing to the one-side generation of kinks on the step. Measuring the angle between the fringe and the edge, we calculated the maximum $\lambda = 18s$. While at the same time, we assumed the minimum value from previous studies as a matter of convenience. Vekilov observed the growth interface of apoferritin crystals using an atomic force microscope (AFM) and found that $\lambda \sim 3.5s$ was almost constant in the wide range of supersaturation [10]. He observed irregularly curved steps on the growth interface of apoferritin crystals, whereas the shape of steps on GI crystals was almost straight: the λ of the GI crystal should be larger than $\lambda \sim 3.5s$. Additionally, the above determination of λ is reasonable since Sleutel et al. observed straight arrays of 4–10 GI tetramers at the edge of two-dimensional GI crystals in nucleation processes on mica substrates using an AFM [11]. With these values and the slopes of ground-based data measured in Fig. 8.2 (●), ν was calculated to be 229.8–1182 s^{-1}. We successfully estimated the vibration frequency of protein molecules on kink incorporation processes on the growth interface of crystals for the first time. The frequency is probably 3–4 orders of magnitude smaller than that in a bulk solution: the frequency of rotational diffusion of lysozymes in the solution was estimated at about 10^7 s^{-1} from dielectric relaxation spectra of lysozyme aqueous solution [12]. The smaller frequency of this study would be due to the strong attractive force between a GI tetramer with the growth interface of a GI crystal. As a similar example, surface diffusion coefficients of lysozyme on the growth interfaces of crystals were known to be 4–5 orders of magnitude smaller than its bulk diffusion constants [13]. This supports our results on the vibration frequency of protein molecules, since surface diffusion coefficients of molecules are usually proportional to their vibration frequencies [14].

8.3 "Peeling Off" Processes Induced by the Adsorption of a Particle from Bulk Dispersion of Particles on a Particle on a Step Edge of the Growth Interface of a Colloidal Crystal [6]

8.3.1 Materials

In this study, we used polystyrene particles (Thermo Fischer Scientific, Waltham, MA, USA, 5065A (diameter $d = 0.65$ μm)) without further modification. We used sodium polyacrylate (Kishida Chemical, Osaka, Japan, molecular weight Mw ~2,070,000–6,210,000) as the crystallizing agent. Polystyrene particles (volume fraction $\varphi = 0.006$) and sodium polyacrylate (2.5×10^{-4} g mL^{-1}) were dissolved into ultrapure water (Millipore, resistivity is larger than 18.2 MΩ·cm) just before the experiments were started.

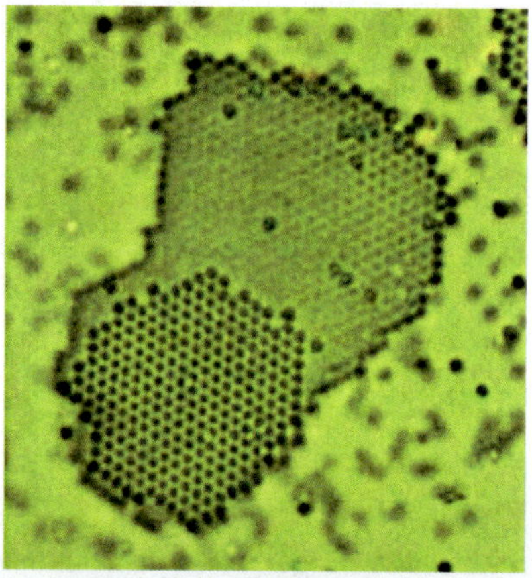

Fig. 8.3 A colloidal crystal with three layers of regular periodic arrays of polystyrene particles. We could clearly resolve individual particles at the particle level. The focal plane is adjusted at the third layer in this picture. Reprinted with permission from the reference [6]. Copyright 2016 MDPI

8.3.2 In Situ Observation of the Growth Interface of Colloidal Crystals Formed with Depletion Attraction Between Polystyrene Particles

We observed growth interfaces of colloidal crystals by optical microscopy using a 100 × oil immersion objective (Fig. 8.3). The first three layers and more were easily resolved at each particle level. Particles on a layer migrated between one specific lattice position and another discretely, while those in bulk water suspension diffused via free Brownian motion simultaneously. We also confirmed straight-shaped steps at the edges of the layer. From the viewpoint of the periodic bond chain (PBC) theory [15], the straightness of steps often reflects the strength of attractive forces between particles at the edge of the steps. On the contrary, the roughness of the steps should be one of the most important variables for roughening transitions of steps. We can solve these problems in real time and analyze the dynamic incorporation processes of particles into the steps using the present experimental apparatuses.

8.3.3 Desorption of Particles from Steps by Attractive Interactions from Other Particles

We found a novel desorption mechanism of a particle at the edge of a step. Desorption proceeded when another particle stuck to the particle at the edge as if the stuck particle peeled the edge particle off the step (Fig. 8.4). Although no textbook of crystal growth has described such a phenomenon, this "peeling off" process is naturally acceptable owing to the attractive forces between particles. We observed not only

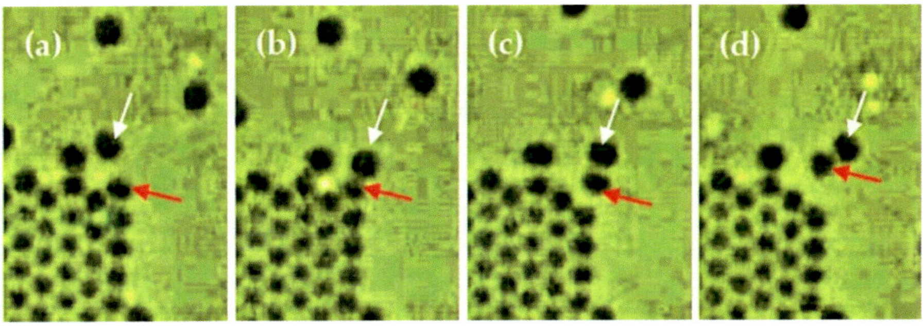

Fig. 8.4 Frame-by-frame advance of a "peeling off" process induced by sticking of a particle from surrounding dispersion onto a particle on a step. **a** A particle on a terrace of a crystal approaches a top layer; **b** The particle sticks to an edge particle of the top layer; **c** The edge particle is separated from the layer by sticking the adsorbed particle; **d** The particle pair is entirely peeled the particle of the top layer. Reprinted with permission from the reference [6]. Copyright 2016 MDPI

this "peeling off" process but also cooperative rearrangements in clusters of particles around steps. Lattice-vibration-like cooperative movements of particles in the terrace of the colloidal crystal were also observed. This is probably due to the attractive interactions between particles inside the crystal. More detailed observation of these cooperative phenomena of particles will be probably useful to discuss the elementary processes on interface-reaction-limited processes of atomic or molecular crystals deeply.

8.4 Conclusions

In this review, we presented several examples of the effectiveness of in-situ observations of crystal growth processes. The key points are summarized as follows.

1. V_{lateral} in space and V_{step} on the earth were almost the same up to $C - C_{\mathrm{e}} = 10\ \mathrm{mg\ mL^{-1}}$. Similar crystal qualities would be realized with similar step velocities: we don't have to grow protein crystals under microgravity as far as we use highly purified proteins.
2. Vibration frequency ν of GI tetramers at kink sites was for the first time estimated to be 229.8–1182 $\mathrm{s^{-1}}$ from the slope of V_{step} with respect to $C - C_{\mathrm{e}}$.
3. Desorption of a particle at the edge of a step induced by the adsorption of an additional particle to the edge particle was newly observed as if the additional adsorbed particles peeled the edge particle off the step [6].

References

1. Nakamura, S.: GaN growth using GaN buffer layer. Jpn. J. Appl. Phys. **30**, L1705–L1707 (1991)
2. Kim, C.U., Lew, W., Williams, M.A., Liu, H., Zhang, L., Swaminathan, S., Bischofberger, N., Chen, M.S., Mendel, D.B., Tai, C.Y., Laver, W.G., Stevens, R.C.: Influenza neuraminidase inhibitors possessing a novel hydrophobic interaction in the enzyme active site: design, synthesis, and structural analysis of carbocyclic sialic acid analogues with potent anti-influenza activity. J. Am. Chem. Soc. **119**, 681–690 (1997)
3. Davis, A.M., Teague, S.J., Kleywegt, G.J.: Application and limitation of x-ray crystallographic data in structure-based ligand and drug design. Angew. Chem. Int. Ed. **42**, 2718–2736 (2003)
4. Suzuki, Y., Tsukamoto, K., Yoshizaki, I., Miura, H., Fujiwara, T.: First direct observation of impurity effects on the growth rate of tetragonal lysozyme crystals under microgravity as measured by interferometry. Cryst. Growth Des. **15**, 4787–4794 (2015)
5. Suzuki, Y., Ninomiya, A., Fukuyama, S., Shimaoka, T., Nagai, M., Inaka, K., Yanagiya, S., Sone, T., Wachi, S., Kawaguchi, S., Arai, Y., Tsukamoto, K.: Highly purified glucose isomerase crystals under microgravity conditions grow as fast as those on the ground do. Cryst. Growth Des. **22**, 7074–7078 (2022)
6. Suzuki, Y., Hattori, Y., Nozawa, J., Uda, S., Toyotama, A., Yamanaka, J.: Adsorption, desorption, surface diffusion, lattice defect formation, and kink incorporation processes of particles on growth interfaces of colloidal crystals with attractive interactions. Crystals **6**, 80 (2016)
7. Koizumi, H., Tachibana, M., Yoshizaki, I., Fukuyama, S., Tsukamoto, K., Suzuki, Y., Uda, S., Kojima, K.: Dislocations in high-quality glucose isomerase crystals grown from seed crystals. Cryst. Growth Des. **14**, 5111–5116 (2014)
8. Suzuki, Y., Sazaki, G., Visuri, K., Tamura, K., Nakajima, K., Yanagiya, S.: Significant decrease in the solubility of glucose isomerase crystals under high pressure. Cryst. Growth Des. **2**, 321–324 (2002)
9. Yoshizaki, I., Tsukamoto, K., Yamazaki, T., Murayama, K., Oshi, K., Fukuyama, S., Shimaoka, T., Suzuki, Y., Tachibana, M.: Growth rate measurements of lysozyme crystals under microgravity conditions by laser interferometry. Rev. Sci. Instrum. **84**, 103707 (2013)
10. Vekilov, P.G.: What determines the rate of growth of crystals from solution? Cryst. Growth Des. **7**, 2796–2810 (2007)
11. Sleutel, M., Lutsko, J., Van Driessche, A.E.S., Durán- Olivencia, M.A., Maes, D.: Observing classical nucleation theory at work by monitoring phase transitions with molecular precision. Nat. Commun. **5**, 5598 (2014)
12. Cametti, C., Marchetti, S., Gambi, C.M.C., Onori, G.: Dielectric relaxation spectroscopy of lysozyme aqueous solutions: analysis of the δ-dispersion and the contribution of the hydration water. J. Phys. Chem. B **115**, 7144–7153 (2011)
13. Sazaki, G., Okada, M., Matsui, T., Watanabe, T., Higuchi, H., Tsukamoto, K., Nakajima, K.: Single-molecule visualization of diffusion at the solution-crystal interface. Cryst. Growth Des. **8**, 2024–2031 (2008)
14. Burton, W.K., Cabrera, N., Frank, F.C.: The growth of crystals and the equilibrium structure of their surfaces. Phil. Trans. R. Soc. A **243**, 299–358 (1951)
15. Hartman, P.: Crystal growth: An introduction. North-Holland, Amsterdam (1973)

Chapter 9
Approach for Achieving Effective Photocatalytic Activity Under Visible Light of WO_{3-x}/SnO_2 Produced by Laser Ablation Method

Vinayak Shinde, Tetsuro Katayama, Yasuyuki Maeda, Satoshi Sugano, Akihiro Furube, and Pankaj Koinkar

Abstract This work successfully prepared tungsten suboxide/tin oxide (WO_{3-x}/SnO_2) composites via Nd:YAG laser ablation method. The synthesized WO_{3-x}/SnO_2 binary composite was used as a photocatalyst for the decomposition. The binary composites exhibited higher photocatalytic performance under LED light irradiation than pristineWO_{3-x} and SnO_2. The obtained materials were characterized by microscopic and spectroscopic techniques, namely, energy dispersive X-ray spectroscopy (EDS), scanning electron microscopy (SEM), and UV–Vis spectroscopy. UV–Vis spectroscopy provides evidence that tungsten suboxide and crystalline SnO_2 nanoparticles have formed, and the visible area absorption of the composite material indicates suitability for photocatalytic activity. The SEM images of the WO_{3-x}/SnO_2 revealed that SnO_2 octahedron crystals were deposited on the petal-like structure of WO_{3-x}. Furthermore, elemental mapping was performed using EDS to confirm the presence of WO_{3-x}, SnO_2, and WO_{3-x}/SnO_2 binary composite. XRD peak positions matching with JCPDS standards confirmed the formation of the binary composite. The inclusion of SnO_2, which may increase the separation efficiency of electron–hole pairs, is directly responsible for the improved photocatalytic performance of the WO_{3-x}/SnO_2 composite. After 60 min of illumination, 84.4% of the MB was removed as compared to 41.6% and 25.4% exhibited by pristine SnO_2 and WO_{3-x}, respectively.

V. Shinde · Y. Maeda
Department of Optical Science, Graduate School of Technology, Industrial and Social Sciences, Tokushima University, 2-1 Minamijosanjima Cho, Tokushima 770 8506, Japan

T. Katayama · A. Furube · P. Koinkar (✉)
Institute of Post-LED Photonics and Graduate School of Technology, Industrial and Social Sciences, Tokushima University, 2-1 Minamijosanjima Cho, Tokushima 770 8506, Japan
e-mail: koinkar@tokushima-u.ac.jp

S. Sugano
Tokushima University, 2-1 Minamijosanjima Cho, Tokushima 7708506, Japan

R. Murakami et al. (eds.), *The 3rd International Conference on Nanomaterials and Advanced Composites*, Springer Proceedings in Physics 298,
https://doi.org/10.1007/978-981-99-7153-4_9

For the purpose of a high photocatalytic activity toward MB photodecomposition, WO_{3-x}-based photocatalyst is one of the most promising candidates. Hence, this research study shows that the WO_{3-x}/SnO_2 composite has a significant potential for use in heterogeneous photocatalysis.

9.1 Introduction

In the recent past, various materials have been used in the photocatalytic process of degrading organic pollutants in sewage. Mainly semiconductor structures are used for this purpose and their photochemical nature drastically changes with a composite form called heterostructure. Two-dimensional transition metal dichalcogenides (2D TMDs) are becoming game changers as photocatalysts and are attracting the greater interest of scientists and researchers in decomposing environmentally hazardous chemicals and pollutants because of their significant response in photocatalytic activities [1] and remarkable electronic and optical properties with better stabilities. One of the representative transition metal materials, WS_2, has a layered two-dimensional structure in nature in bulk form with an indirect band gap of 1.4 eV. When it is scaled down from bulk to nanosize and forms a sheet with a few layers, a 2.0 eV direct bandgap with increased optical absorption in the visible region is observed. WS_2 is a leading candidate for creating a base for visible light-active photocatalysis because of its low band gap and stable chemical features. On the other hand, tin oxide (SnO_2) use is increasing with more research carried out for it as a photocatalyst due to its stability, better photoactivity, low cost, and eco-friendly nature. SnO_2 has a large band gap (3.2 eV) which restricts its light absorption in the UV region, so applying SnO_2 as ideal photocatalytic material gets limitations. Another restraint towards getting high photocatalytic efficiency was poor charge separation efficiency.

The tungsten suboxide with nanopetal structure with sizes 300–500 nm was also well suited for the substrate of the composite. With reference to earlier research, the separation rate of the photogenerated electron–hole pairs can be enhanced notably because of the various energy band levels present in the composite material. The combination of distinct band gap energies with the compound of transition metals may also cause a significant expansion of the spectral response.

In this study, we have synthesized SnO_2/WO_{3-x} heterostructure photocatalyst with octahedron crystals with nanopetal structures via a simple laser ablation method. In comparison with other synthesis methods like the hydrothermal or chemical vapor deposition methods, laser ablation is relatively clean, fast, and tuneable of parameters such as ablation time, laser power, frequency, and focal distance, making it the most effective for high yield and good stability in synthesized materials. Our efforts were made to reveal the formation mechanism of WO_{3-x} nanopetal-like structure from bulk WS_2 as well as the photocatalytic properties of synthesized crystalline SnO_2 and the combined SnO_2 and WO_{3-x} composites. The combination between SnO_2 and WO_{3-x} reveals remarkable improvements in photocatalytic activity. The high chemical stability of SnO_2/WO_{3-x} composite photocatalysts suggests the possible realization and use of it in industrial applications related to environmental safety.

9.2 Experimental

9.2.1 Preparation of WO_{3-x}

In this experiment, we used tungsten disulfide (WS_2) powder (10 mg) as a bulk material and added it in a similar amount of distilled water (10 ml) and ethanol (10 ml) followed by probe sonication using VIOLAMO Sonicstar 85 for 10 min before exposing it to nanosecond laser ablation. The bulk WS_2 powder purchased from Sigma-Aldrich with an average sheet size of about 4–6 μm was used for this experiment. The sample underwent constant magnetic stirring while undergoing ns laser ablation with the second harmonic of a Nd:YAG laser (532 nm, repetition frequency 10 Hz, 600 mW, 10 ns) for 20 min. The non-ablated and ablated samples were characterized using UV–Vis Spectroscopy (Hitachi, U-2010), FE-Scanning Electron Microscopy (Hitachi, S4700), and EDS (JEOL, JSM-6510A).

9.2.2 Preparation of SnO_2

The bulk SnO powder from Sigma-Aldrich with an average sheet size of about a few μm was used for this experiment. The SnO_2 octahedron crystal-like structure was fabricated using tin oxide powder (SnO) (10 mg) as a bulk material by adding it to ethanol (20 ml), and the probe sonicated it for 10 min using VIOLAMO Sonicstar 85 before irradiating with Nd:YAG laser ablation. After that, the sample underwent ns laser ablation for 20 min with constant magnetic stirring. Second harmonic Nd:YAG laser of the second harmonic having a wavelength of 532 nm with the power of 600 mW was used for irradiation. Characterization of the sample before and after ablation was done using UV–Vis Spectroscopy, FE-Scanning Electron Microscopy, and EDS.

9.2.3 Preparation of SnO_2/WO_{3-x}

The laser-ablated WO_{3-x} and SnO_2 were then dried using the vapor evaporation method and the octahedron phase blue powder of WO_{3-x} and white-yellowish SnO_2 powder were obtained. These two powdered samples were mixed in distilled water of (10 ml) with a proportion of 1 mg each and bath sonicated for 1 h. Then characterizations were performed using UV–Vis Spectroscopy, FE-Scanning Electron Microscopy, and EDS.

9.2.4 Photocatalysis Experiments

Photodegradation of methylene blue (MB) was used to investigate the photocatalytic activity of all prepared samples, namely WO_{3-x}, SnO_2, and WO_{3-x}/SnO_2 composite. The photocatalytic activity of these solvents was evaluated by combining them with an aqueous solution of MB while keeping the mixture under continuous magnetic stirring. The mixture was then exposed to 210 mW of LED light using a UV cut filter (330 nm) at a distance of 50 cm. The absorption measurements were taken for dark time (−30 min), at time zero, and then up to 60 min with a time interval of 10 min to observe the step-by-step photodegradation of MB.

9.3 Results and Discussion

9.3.1 UV–Vis Spectroscopy

Figure 9.1 shows the UV–Vis spectra of sub-oxide WO_{3-x}, SnO_2, and the binary composite of WO_{3-x}/SnO_2. Semiconductors were known for their excellent optical properties and hence found usage in optoelectronic materials [2]. In tungsten suboxide spectra, the small hump structure was around at ~270 nm, suggesting the bulk WS_2 was modified into (WO_{3-x}) tungsten sub-oxide [3]. After ablation, a rise in absorption indicates the exfoliation of tungsten disulfide microsheets in nanopetals. The SnO_2 nanocrystals showed a broad and strong absorption around 247 nm indicating the crystalline formation of as-synthesized SnO_2 [4]. The UV–Vis of binary composite WO_{3-x}/SnO_2 showed structureless light absorption in visible or near visible regions suggesting the photocatalytic property of prepared composite material [5].

9.3.2 FE SEM Analysis

FE SEM analysis is typically used to understand the surface morphology and capture the micro/nanostructure of the material. As Fig. 9.2 shows FESEM images of (a) bulk WS_2, (b) WO_{3-x}, (c) bulk SnO, (d) ablated SnO_2, (e) WO_{3-x}/SnO_2 and their morphological structures are easily distinguishable before and after ablation for all the prepared samples. Figure 9.2a shows a few micrometer sheets of bulk WS_2,

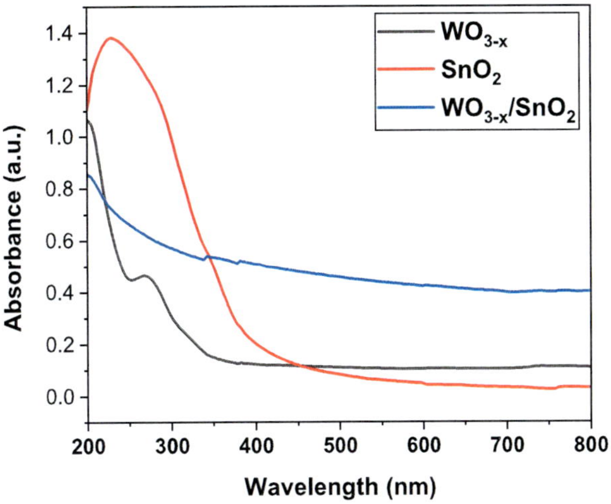

Fig. 9.1 UV–visible spectra of WO_{3-x}, SnO_2 and WO_{3-x}/SnO_2

followed by (b) showing the petal-like morphological and evenly spread structure in tungsten suboxide formation with almost 70% area covered. SnO_2 is an n-type versatile semiconductor material with a crystalline nature that shows agglomeration of bulk SnO with the size of ~1 μm sheets in Fig. 9.2c. Figure 9.2d shows the very well-grown and crystalline form of the octahedron shape of SnO_2 after ablation, with sizes ranging from 100 to 150 nm. The formation of six bonds is more favorable than four as crystal field stabilization energy is always higher for octahedral than tetrahedral crystals [6]. Figure 9.2e demonstrates the uniform formation of the WO_{3-x}/SnO_2 composite while keeping the entire morphology intact after the laser ablation. This indicates that the surface morphological features that have been intact and will complement each other, resulting in better photocatalytic properties.

9.3.3 *EDS Analysis*

EDS is used to determine the chemical composition present in a sample and to estimate its relative abundance. Figure 9.3a shows the EDS spectra of bulk WS_2, and Fig. 9.3b suggests elemental mapping of WO_{3-x} formation after ns laser ablation. As per the graph, it is found that WO_{3-x} lies between $0 < x < 0.4$, where x is 0.22, which confirms the formation of tungsten suboxide. Figure 9.3c shows the EDS spectra of bulk SnO, and Fig. 9.3d shows the elemental composition of the synthesized SnO_2 octahedron crystal structure. This graph is auto-generated by built-in software with JEOL JSM 6510A. From this spectrum, noticeable and dominant peaks belong to tin and oxygen, which confirms that the as-synthesized octahedron nanocrystals were made of these two elements. We can say that the as-synthesized powder is in

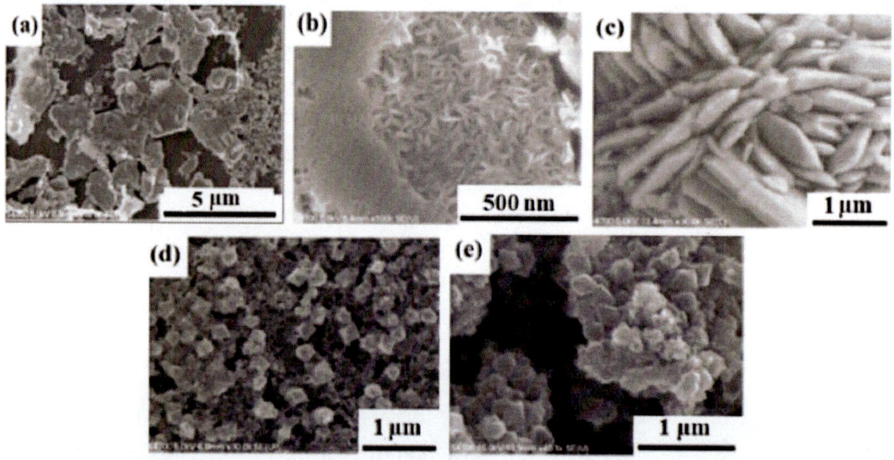

Fig. 9.2 FESEM images of (**a**) bulk WS_2, (**b**)_WO_{3-x}, (**c**) bulk SnO, (**d**) ablated SnO_2, and (**e**) WO_{3-x}/SnO_2

pure form due to the presence of these two peaks, while other peaks are negligible. Figure 9.3e demonstrates the presence of all elements in the WO_{3-x}/SnO_2 composite.

9.3.4 XRD Analysis

XRD pattern of WO_{3-x}/SnO_2 binary nanocomposite is shown in Fig. 9.4. The octahedron crystals and nanopetal flake-like structures of the prepared composite will be strongly suggested by sharp diffraction peaks. The diffraction peaks of the binary composite can be seen in the WO_{3-x} nanopetal-like flake structure (JCPDS card No. 33–1387), and SnO2 octahedron crystal (JCPDS card No. 41–1445) [7]. The XRD peaks at 2θ values of 26.94°, 33.35°, 37.54°, and 52.08° reveal the formation of octahedron crystals of SnO_2. Also, XRD peaks at 2θ values of 30.26°, 34.5°, 51.24°, and 57.73° reveal the formation of nanopetal flake-like structure of WO_{3-x} [8].

9.3.5 Photocatalysis

The photocatalytic activity was calculated with the help of a UV spectroscopy, and the resulting absorption spectra for the degradation of MB are presented in Fig. 9.5a, b and c for WO_{3-x} nanopetals, SnO_2 octahedron crystals, and WO_{3-x}/SnO_2 composite, respectively.

As shown in Fig. 9.5, the different materials used affect the degradation efficiency of MB with UV–Vis light irradiation (> 330 nm). The photodegradation efficiency of all materials was calculated using below formula:

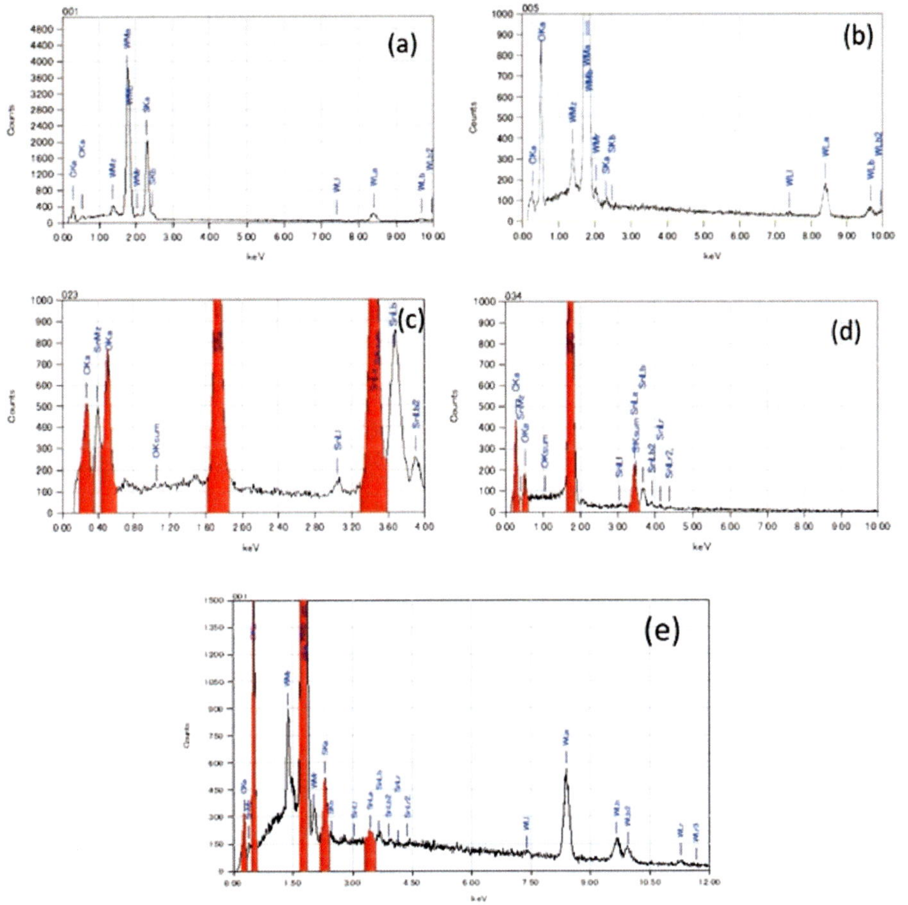

Fig. 9.3 EDS spectra of (**a**) bulk WS_2, (**b**) WO_{3-x}, (**c**) bulk SnO, (**d**) ablated SnO_2, and (**e**) WO_{3-x}/SnO_2

Fig. 9.4 XRD spectrum of WO_{3-x}/SnO_2 composite

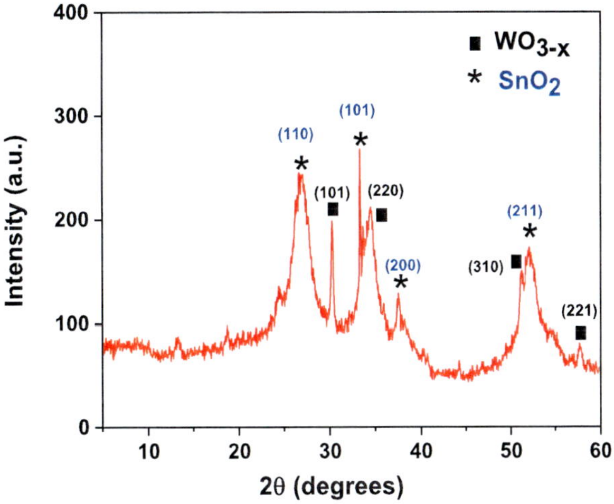

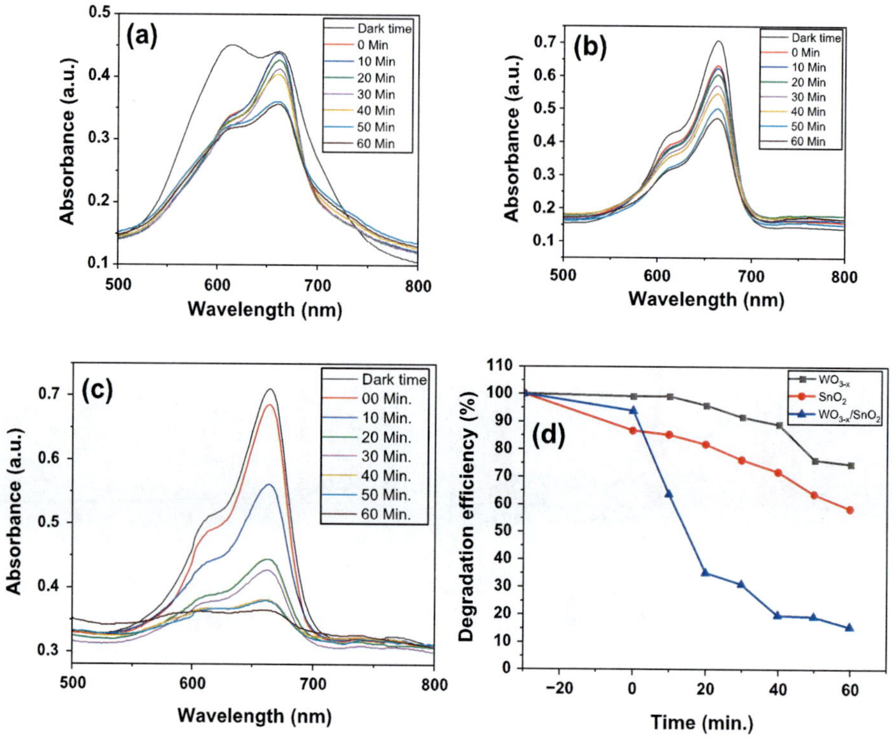

Fig. 9.5 UV–vis spectra of (**a**) WO_{3-x}, (**b**) SnO_2 octahedron crystals, and (**c**) WO_{3-x}/SnO_2 binary composite regarding the photocatalytic degradation of MB also (**d**) degradation efficiency in percentages of WO_{3-x}, SnO_2, and WO_{3-x}/SnO_2

$$Degradation\ efficiency(\%) = \frac{A_0 - At}{A_0} \times 100\%$$

Here, A_0 and At denote MB absorption intensity at time zero and 60 min., respectively.

The as-synthesized SnO_2 octahedron crystals exhibited a degradation efficiency of 41.6% after 60 min. as depicted in Fig. 9.5b. In comparison, Fig. 9.5a WO_{3-x} nanopetal-like structure showed a degradation efficiency of 25.4% after the same duration of 60 min. The degradation efficiency of binary composite WO_{3-x}/SnO_2 is 84.4% as per Fig. 9.5c. In comparison with as-synthesized tungsten suboxide and SnO_2, the prepared binary composite shows better degradation potential. Figure 9.5d shows the comparative and graphical representation of the degradation efficiency of prepared samples. Different photocatalytic activity may result from various parameters, including band gap characteristics, nanopetals or crystals, morphological structure, photochemical stability, and crystal defect [9]. Dye molecules break down naturally when exposed to oxygen in the air and UV radiation from the sun [10, 11]. Since this degrading activity is typically slow, involving a semiconductor material as a photocatalyst will significantly speed up the process. Similar studies have shown how semiconductor materials exhibited non-considerable photocatalytic decomposition of dyes in the dark phase. Heterogeneous photocatalysis is often used to degrade hazardous molecular compounds. This suggests that material like WO_{3-x}/SnO_2 absorbs the required energy from light. The light energy absorbed by the material exposed promotes the movement of electrons from the valence band (VB) to the conduction band (CB) of the photocatalytic material. This causes a responsive electron–hole pair to form, moving toward the water-semiconductor interface, where electrons are transferred between nearby species. Thus, the result of this interaction is the removal of contaminants from the medium [12]. So, this study also confirms that photocatalytic degradation hardly proceeds without light [13].

9.4 Conclusion

In this study, we effectively produced the WO_{3-x}/SnO_2 composite using ns laser ablation in the liquid phase, and its photocatalytic performance was evaluated in respect to degradation of MB. The UV–Vis spectra of all prepared samples suggest their suitability for photocatalytic activity, which was further confirmed using photocatalytic degradation of MB. EDS spectra indicate the formation of tungsten suboxide from bulk WS_2 and nanocrystal octahedron structure from bulk SnO after ns ablation. It also confirms the formation of compound WO_{3-x}/SnO_2 with elemental mapping. SEM images show a uniform nanopetal-like structure in WO_{3-x} and octahedron crystal in as-synthesized SnO_2. The combined morphology of these two was observed in binary composite WO_{3-x}/SnO_2. The intensity of peaks is higher in XRD due to good crystallinity. The photocatalytic activity was checked and compared with the help of photodegradation of MB using WO_{3-x}, SnO_2, and the binary composite. The WO_{3-x}/SnO_2 composite shows the best result with 84.4% degradation efficiency.

Acknowledgements The authors would like to acknowledge the Faculty of Science and Technology, Tokushima University, Japan for financial support through the Advanced Science and Engineering Education Research Project (2022).

References

1. Sang, Y., Zhao, Z., Zhao, M., Hao, P., Leng, Y., Liu, H.: From UV to Near-infrared, WS_2 Nanosheet: A novel photocatalyst for full solar light spectrum photodegradation. Adv. Mater. **27**(2), 363–369 (2015)
2. Hu, X., Song, G., Li, W., Peng, Y., Jiang, L., Xue, Y., Liu, Q., Chen, Z., Hu, J.: Phase-controlled synthesis and photocatalytic properties of SnS, SnS_2 and SnS/SnS_2 heterostructure nanocrystals. Mater. Res. Bull. **48**(6), 2325–2332 (2013)
3. Famili, Z., Dorranian, D., Sari, A. H.: Laser ablation-assisted synthesis of tungsten sub-oxide ($W_{17}O_{47}$) nanoparticles in water: effect of laser fluence. Opt. Quantum Electron. **52**(6), Article number 305, (2020)

4. Adeyemi, J.O., Onwudiwe, D.C.: SnS_2 and SnO_2 Nanoparticles obtained from organotin (IV) Dithiocarbamate complex and their photocatalytic activities on methylene blue. Materials **13**(12), Article number 2766 (2020)
5. Gupta, P., Lapalikar, V., Kundu, R., Balasubramanian, K.: Recent advances in membrane based waste water treatment technology: A review. Energy Environ. Focus. **5**(4), 241–267 (2016)
6. Huang, X., Nan, Z.: Formation of octahedron-shaped $ZnFe_2O_4/SiO_2$ with yolk–shell structure. J. Phys. Chem. Solids, **141**, Article number 109410 (2020)
7. Xue, X.-Y., He, B., Yuan, S., Xing, L.-L., Chen, Z.-H., Ma, C.-H.: SnO_2/WO_3 core–shell nanorods and their high reversible capacity as lithium-ion battery anodes. Nanotechno, **22** (39), Article number 395702, (2011)
8. Garde, A.S.: Electrical and structural properties of WO_3-SnO_2 thick-film resistors prepared by screen printing technique. Res J Recent Sci **4**(1), 55–61 (2015)
9. Zhang, Y.C., Du, Z.N., Li, S.Y., Zhang, M.: Novel synthesis and high visible light photocatalytic activity of SnS_2 nanoflakes from $SnCl_2.2H_2O$ and S powders. Appl. Catal. B: Environ. **95**(1–2), 153–159 (2010)
10. Julkapli, N.M., Bagheri, S., Hamid S.B.A.: Recent advances in heterogeneous photocatalytic decolorization of synthetic dyes. Sci. World J., **2014**. Article number 692307
11. Lam, S.-M., Sin, J.-C., Abdullah, A.Z., Mohamed, A.R.: Degradation of wastewaters containing organic dyes photocatalysed by zinc oxide: a review. Desalin. Water Treat. **41**(1–3), 131–169 (2012)
12. Xie, Y., Zhang, C., Miao, S., Liu, Z., Ding, K., Miao, Z., An, G., Yang, Z.: One-pot synthesis of ZnS/polymer composites in supercritical CO_2–ethanol solution and their applications in degradation of dyes. J. Colloid Interface Sci. **318**(1), 110–115 (2008)
13. Sharma, M., Jain, T., Singh, S., Pandey, O.P.: Photocatalytic degradation of organic dyes under UV-Visible light using capped ZnS nanoparticles. Sol. Energy **86**(1), 626–633 (2012)

Chapter 10
Study on Cellulose Nanofiber Molding by 3D Printing

Yuta Yokota, Antonio Norio Nakagaito, and Hitoshi Takagi

Abstract Cellulose nanofibers (CNF) are obtained by finely unraveling fibers extracted from plants like wood, into nano-sized elements. CNF is lightweight yet has high strength and little thermal expansion. It is known to be five times lighter than steel but more than five times stronger. With the conventional method, it takes a long time to manufacture a molded product from CNF as it requires a long filtration step of the CNF aqueous suspension, which is an obstacle to fabrication. In this experiment, we propose to establish a new molding process with the aim of shortening the time required to manufacture molded products. Specifically, it is a method of connecting a home-made syringe pump to a commercial 3D printer to directly print the CNF suspension into shape. By adopting this method, filtration can be omitted from the conventional molding process. A concentrated CNF suspension is used as the feeding material for the 3D printer, and the syringe pump enables the extrusion of CNF hydrogel into the 3D printer. As CNF suspension has high thixotropic properties, it is expected that shear-thinning allows proper extrusion through the narrow channels. Experiments were conducted with the aim of establishing a method for accelerating the production of CNF-molded products.

10.1 Introduction

Cellulose nanofibers (CNF) are finely nanofibrillated fibrils obtained from sources like wood. Fruit pomace and rice straw can also be brought into this state by chemically and mechanically disentangling the nanofibers. In addition, CNF has features such as being lightweight, showing high strength, little thermal expansion, and high

Y. Yokota
Graduate School of Advanced Technology and Science, Tokushima University, Minamijosanjima-cho 2-1, Tokushima-shi, Tokushima 770-8506, Japan

A. N. Nakagaito (✉) · H. Takagi
Graduate School of Technology, Industrial and Social Sciences, Tokushima University, Minamijosanjima-cho 2-1, Tokushima-shi, Tokushima 770-8506, Japan
e-mail: nakagaito@tokushima-u.ac.jp

R. Murakami et al. (eds.), *The 3rd International Conference on Nanomaterials and Advanced Composites*, Springer Proceedings in Physics 298,
https://doi.org/10.1007/978-981-99-7153-4_10

transparency. It is known to be five times lighter than steel but more than five times stronger [1]. In addition, since it is more recyclable than glass fiber and carbon fiber, the expansion of the market for products using plant-derived CNF would lead to the elimination of fossil resources and the construction of a recycling-oriented social infrastructure.

As already mentioned, CNF can also be extracted from plants other than trees. The cell structure of plants varies greatly depending on the species, ranging from long honeycomb-like wood cells to foam-like polyhedral parenchyma cells [2]. The wood cell wall is composed of overlapping primary wall and three secondary walls, and it cannot be readily fibrillated unless a strong force is applied. On the other hand, the parenchyma cell wall in fruits is composed of a thin single primary wall, being relatively easier to fibrillate than wood. Citrus fruit residues have this advantage as sources of CNF extraction.

After extracting the CNF from the raw material, we start to manufacture the molded product, but the conventional method requires a long time of filtering and drying, which is very time-consuming. The process of preparing the suspension and filtering takes about 1 h per 100 ml of suspension, and this task is unavoidable if the molded product has a certain thickness. In this experiment, for the purpose of shortening the time required to produce molded products, a method of directly printing CNF suspension by connecting a syringe pump to a 3D printer was adopted.

The RepRap project started by Dr. Adrian Bowyer of the UK in 2005 has greatly spread 3D printers, and recently the burden of commercialization has been reduced. The RepRap project is a project to develop a desktop-sized 3D printer as open source, and it is characterized by "self-replication" that can make plastic parts necessary for making another 3D printer by itself. Many 3D printers, such as Prusa and Ender, which are widely used today, have been distributed as open-source hardware as a result of the project [3]. In this way, the introduction of 3D printers became easier, and many studies are now based on them.

If it would be possible to stably produce CNF molded products using a 3D printer, the filtration work would be omitted and then the required time would be shortened. Since CNF suspension is used as the material for the 3D printer in the current study, a syringe pump that enables the extrusion of gel-like material was built and connected to the 3D printer. As CNF has a high thixotropic property, it is expected to work effectively during extrusion from a syringe pump. The purpose of this experiment was to establish a method for stably producing molded products.

10.2 Methods

10.2.1 Connection of Syringe Pump to 3D Printer

The syringe pump and the 3D printer were connected with reference to Leech et al. [4]. The syringe pump was built following the open-source project by Pusch et al. [5] by printing the parts of the pump in polylactic acid with a 3D printer. The printer used was Ender-3 (Shenzen Creality 3D Technology Co., Ltd.), and the slicer software was Ultimaker Cura (Ultimaker BV). The G-code was edited using Repetier-Host (Hot-World GmbH & Co. KG). A flexible tube was attached to the tip of the syringe at one end and to a metal needle with an inner diameter of 1.1 mm to the other. The extruder print head of the 3D printer was replaced by a plastic support for the needle and the needle itself. Figure 10.1 shows the whole printer setup equipped with the syringe pump.

10.2.2 Pulp Extraction

The starting material was sudachi (*Citrus sudachi*) citrus fruit residue after juice extraction. Pulp was extracted by a modified Wise method. Chemical treatments were performed in the order of pigment removal, pectin removal, and hemicellulose removal. To remove pigments, the fruit residue was stirred at 75 °C in 10 g of sodium chlorite and 2 mL of acetic acid added to 1.5 L of distilled water. Equal amounts of sodium chlorite and acetic acid were added every hour and stirred for a total of 3 h. After treatment, the material was washed with water until it became neutral. Next, to remove pectin, the material was added to an aqueous solution of distilled water and hydrochloric acid at 0.18 wt% and treated at 120 °C for 2 h. After treatment, the material was again washed with water until it became neutral. Finally, to remove

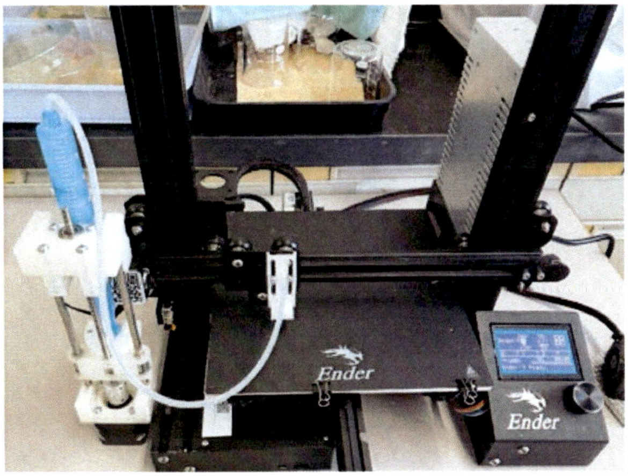

Fig. 10.1 Syringe pump-equipped 3D printer setup

hemicelluloses, the material was treated in 6 wt% potassium hydroxide aqueous solution at 80 °C, stirring for 2 h. After that, the material was washed with water until it became neutral.

10.2.3 Preparation of Suspension

Distilled water was added to the extracted pulp to prepare a suspension with a fiber content of 0.8 wt%. The suspension was placed in a kitchen blender and stirred for 30 min to obtain a CNF suspension. After fibrillation, the CNF suspension was transferred to a beaker and stirred under vacuum for defoaming at 180 rpm for 6 h. The entrapped air was removed from the suspension by stirring under negative pressure. After defoaming, the suspension was vacuum filtered until reaching 3 wt% fiber content.

10.2.4 Fabrication of CNF Moldings

A suspension of 3 wt% fiber content was put into a syringe connected to a 3D printer. The printing file was generated by Ultimaker Cura (Ultimaker BV) and using Repetier-Host (Hot-World GmbH & Co. KG), G-code was edited to conform to printing with the syringe pump. The CNF suspension was printed in layers inside a metal mold and dried at 105 °C for 24 h in an oven dryer with a weight placed on it.

10.2.5 Tensile Test

Dried moldings were cut into 10 mm × 50 mm to prepare test pieces. Specimens were oven-dried at 105 °C for 1 h before testing. Tensile tests were performed with a universal testing machine at a gauge length of 30 mm and cross-head speed of 1 mm/min and tensile strength and Young's modulus were determined.

10.3 Results and Discussion

10.3.1 Effect of CNF Suspension Extrusion During Printing on the Mechanical Properties of Molded Products

The optimal extrusion amount when printing CNF suspension with a 3D printer was investigated. When producing CNF molded products with a 3D printer, the CNF suspension is discharged from the tip of the needle and stacked one layer at a time. If the amount of extrusion was large, the printing time would be short, and if it was small, the printing time would be long. Considering only the printing time, the larger the discharged amount, the better, but we expected that the difference in the discharge amount would affect the mechanical properties of the molded product. Therefore, molded products with the same thickness were fabricated with two different extrusion rates, and their mechanical properties were compared.

The thickness of the printed pieces was set to 20 mm, one by stacking 5 printing layers and the other 12 layers. During drying, a weight was placed on top applying a pressure of about 8 kPa. Both samples reached a thickness of approximately 0.4 mm after drying. Figure 10.2 shows the results of the tensile test. The tensile direction was parallel to the printing direction. The tensile strength of the specimen prepared by laminating five layers was 40 ± 10 MPa, the Young's modulus 3.6 ± 1.0 GPa, and the fracture strain 1.0 ± 0.2%. On the other hand, the tensile strength, modulus, and fracture strain of the specimen prepared by printing 12 layers were 70 ± 11 MPa, 6.2 ± 0.1 GPa, and 1.0 ± 0.2%, respectively.

The difference in strength between the two is thought to be due to the fact that the number of voids inside the test piece decreased in the 12-layer test piece compared to the 5-layer test piece. The densities of both were 0.679 g/cm^3 for the 5-layered specimen and 0.937 g/cm^3 for the 12-layered specimen. The smaller amount of extrusion per layer stacks thinner filaments of material resulting in smaller interlayer voids and higher density.

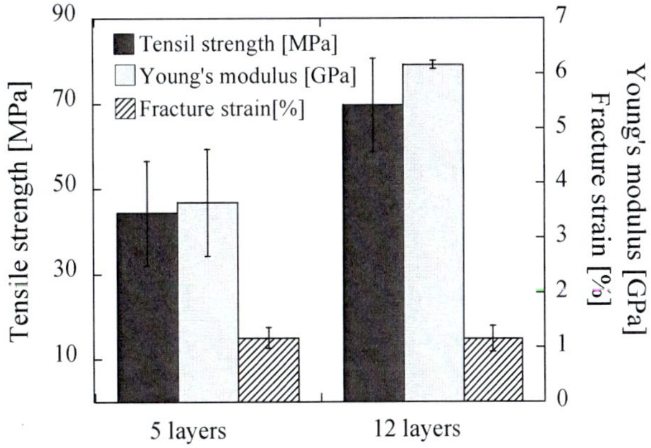

Fig. 10.2 Tensile properties of molded products by stacking 5 and 12 printed layers of CNF

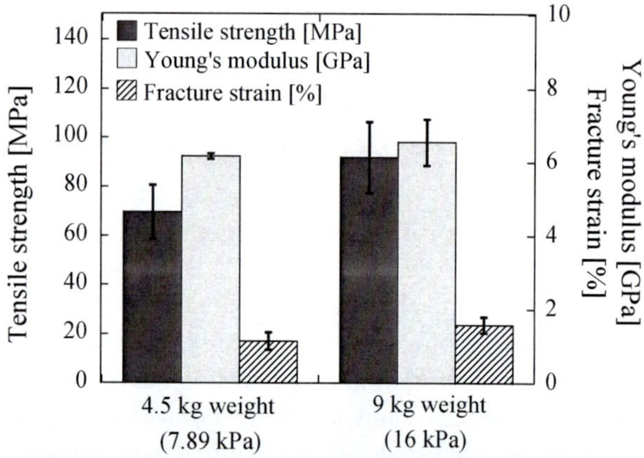

Fig. 10.3 Tensile properties of molded products dried under different pressures

10.3.2 Effect of Drying Pressure on Mechanical Properties of the Molded Product

We investigated the effect of pressure applied during drying of printed CNF suspension on the mechanical properties of molded products. We expected that the higher pressure applied would increase the density of the molded product after drying, delivering better mechanical properties.

Keeping the amount of extrusion the same, the pressure applied during drying was set to approximately 8 kPa and 16 kPa for comparison. Figure 10.3 shows the results of the tensile test. The tensile direction was parallel to the printing direction. The tensile strength of the specimen prepared by applying a pressure of 8 kPa was 70 $\pm$ 11 MPa, the Young's modulus was 6.2 $\pm$ 0.1 GPa, and fracture strain was 1.0 $\pm$ 0.2%. On the other hand, the tensile strength of the specimen prepared by applying pressure of 16 kPa was 90 $\pm$ 10 MPa, the Young's modulus was 6.5 $\pm$ 1.0 GPa, and fracture strain was 2.0 $\pm$ 0.2%. Comparing the densities of the two, the molded product with a pressure of 8 kPa was 0.679 g/cm^3 and the one with a pressure of 16 kPa was 1.24 g/cm^3. An improvement in properties was observed with increasing applied pressure.

10.3.3 Effect of Printing Direction of CNF Suspension on the Mechanical Properties of Molded Parts

We investigated if the printing direction of the layers and the tensile direction of tensile test affects the mechanical properties of the molded parts. Moldings manufactured under the same conditions were tested by changing the tensile direction. One was tested so that the printing direction and the tensile direction were parallel, and the other was tested so that the printing direction and the tensile direction were orthogonal. Figure 10.4 shows the results of the tensile test. The tensile strength of

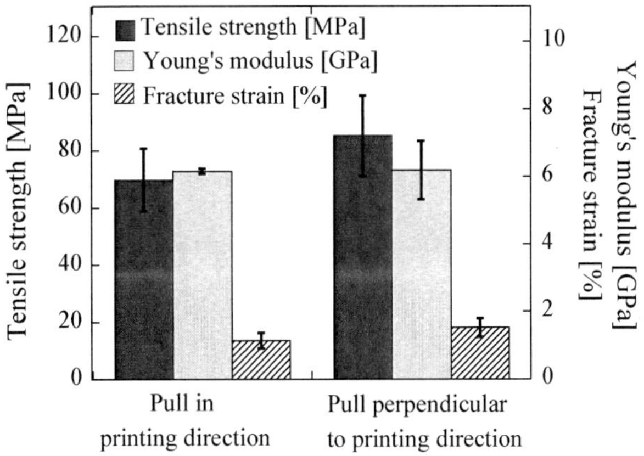

Fig. 10.4 Tensile properties of molded products parallel and perpendicular to printing direction

the specimen tested with the tensile direction parallel to the printing direction was 70 ± 11 MPa, the Young's modulus was 6.2 ± 0.1 GPa, and the fracture strain was 1.0 ± 0.2%. On the other hand, the tensile strength of the specimen tested with the printing direction orthogonal to the tensile direction was 80 ± 10 MPa, the Young's modulus 6.2 ± 1.0 GPa, and fracture strain 2.0 ± 0.3%. The difference was not significant.

Materials made of plant fibers such as wood exhibit remarkable anisotropy with strong tensile strength in the fiber direction. However, the present printed CNF parts do not show such anisotropy. This is probably due to the wide extrusion width during printing. The extruded CNF hydrogel width was about 4 mm, so it is expected that the extrusion width must be narrower to obtain CNF orientation.

10.4 Conclusion

A syringe pump was built and attached to a 3D printer and molded products were fabricated by printing CNF suspensions. From the results of this study, it was found that the density and test direction of the test piece had an effect on the mechanical properties. Since the mechanical properties also showed good results with the increase in density, we will try to further increase the density in the future. As the investigation of the test direction is still incomplete, we would like to clarify it in future investigations. In addition, we would like to investigate if the strength can be maintained by increasing the thickness of the molded product.

It is difficult to control the shrinkage during drying in sample fabrication, and at present, a stable molded product cannot be prepared unless a weight is placed on it. By using a 3D printer, it would be nice to be able to control the shape, but at this stage, it has not reached that point.

References

1. Yano, H.: Resources of Japan, the land of trees. Shinrin Kagaku. **81**, 3–6. The Japanese Forest Society (2017) (in Japanese)
2. Gibson, L.J.: The hierarchical structure and mechanics of plant materials. J. R. Soc. Interface **9**(76), 2749–2766 (2012)
3. All3DP Homepage. https://all3dp.com/history-of-the-reprap-project/. Last accessed 15 Oct 2021
4. Leech, D.J., Lightfoot, S., Huson, D., Stratakos, A.: Low-cost, modular modification to a desktop 3D printer for general purpose gel/Paste extrusion & direct ink writing. (2021). bioRχiv preprint. https://doi.org/10.1101/2021.03.10.434735
5. Pusch, K.: Large volume syringe pump extruder for desktop 3D printers. HardwareX **3**, 49–61 (2018)